Lecture Notes in Artificial I

Subseries of Lecture Notes in Compute

Edited by J. G. Carbonell and J. Siekma

Lecture Notes in Computer Science

Edited by G. Goos, J. Hartmanis and J. van Leeuwen

Springer

Berlin
Heidelberg
New York
Barcelona
Hong Kong
London
Milan
Paris
Singapore
Tokyo

Joachim Paul Walser

Integer Optimization by Local Search

A Domain-Independent Approach

Foreword by Henry Kautz

Springer

Series Editors

Jaime G. Carbonell, Carnegie Mellon University, Pittsburgh, PA, USA
Jörg Siekmann, University of Saarland, Saarbrücken, Germany

Author

Joachim Paul Walser
12 Technologies, Airway Park
Lozenberg 23, B-1932 Sint-Stevens-Woluwe, Belgium
E-mail: walser@i2.com

Cataloging-in-Publication data applied for

Die Deutsche Bibliothek - CIP-Einheitsaufnahme

Walser, Joachim Paul:
Integer optimization by local search : a domain independent
approach / Joachim Paul Walser. - Berlin ; Heidelberg ; New York ;
Barcelona ; Hong Kong ; London ; Milan ; Paris ; Singapore ; Tokyo
: Springer, 1999
   (Lecture notes in computer science ; Vol. 1637 : Lecture notes in
   artificial intelligence)
   ISBN 3-540-66367-3

CR Subject Classification (1998): I.2.8, F.2, G.2, I.2

ISBN 3-540-66367-3 Springer-Verlag Berlin Heidelberg New York

© Springer-Verlag Berlin Heidelberg 1999
Printed in Germany

Typesetting: Camera-ready by author
SPIN 10703430   06/3142 – 5 4 3 2 1 0   Printed on acid-free paper

To my mother, Irene

Foreword

An encouraging trend in the field of artificial intelligence in the past few years has been its growing interaction with the area of operations research. AI has traditionally concentrated on problems of logical inference and satisfiability, or in other words, Boolean feasibility problems. OR, by contrast, has mainly focused on problems of linear optimization. Many significant real world tasks share aspects of both kinds of problems, and there is therefore much interest in integrating and expanding the techniques that have been developed in each field. This monograph by Dr. Joachim Paul Walser is thus particularly timely and significant. He developes new algorithms and systems for applying discrete local search techniques that were originally developed for Boolean feasibility problems to a broad class of integer linear optimization problems. He demonstrates that his general, domain-independent solver can be competitive with specialized algorithms on hard realistic problems, and can often far outperform other state-of-the-art domain-independent solvers. Throughout the monograph Dr. Walser draws connections to classic techniques from OR and AI, and demonstrates how different approaches (such as local search and linear relaxations) can be combined to solve relevant problems in integer programming.

The contributions of this monograph can be placed in two groups: First, there is the specific algorithmic work, including extending the "walksat" algorithm from Boolean to integer constraints and creating a new local search strategy for over-constrained problems. The resulting WSAT(OIP) system represents the distillation of a careful and deep search through the space of possible designs, and its elegance, and empirical success are remarkable. To take one specific example: walksat's performance is known to degrade in the presence of constraints containing many variables. However, WSAT(OIP) can efficiently handle the very long constraints (over 300 variables per constraint) that arise in many real-world domains.

Second, the book presents a series of well-chosen empirical evaluations. The different cases represent a spectrum of different kinds of problems: feasibility versus optimization, loosely-constrained versus highly constrained, and 0/1 versus integer valued variables. Some of the most significant results Dr. Walser reports are on problems of capacitated production planning. These are very large, hard industrial problems, which are (unfortunately) all too rare in academic research. Dr. Walser demonstrates that WSAT(OIP) can find solutions that are much closer to optimal than could be found by any competing approach. The empirical evaluation shows, for the first time, that a completely general local search engine (as opposed to domain-specific local search algorithms) can efficiently find optimal or near-optimal solutions to a broad range of real-world problems, and thus complement established systematic problem-solving frameworks such as integer prorgramming branch-and-bound.

In short, this material is mandatory reading for researchers in AI who are seriously concerned with solving hard combinatorial search and optimization problems, as well as those researchers in OR who want to see the best that AI has to offer. The clarity and breadth of the presentation also makes this book an excellent choice for reading material in a graduate or advanced undergraduate seminar in either AI or OR.

February 1999   HENRY A. KAUTZ
Florham Park, New Jersey, USA

Preface

Software to support complex planning decisions is becoming a vital factor for competitiveness, driven by the increasing availability of organizational data in modern enterprise information systems. Decision support software can reduce manufacturing costs, increase organizational efficiency, and deliver solutions to complex resource allocation problems – by building on effective optimization algorithms.

Integer optimization covers a variety of practically important optimization problems, including production planning, timetabling, VLSI circuit design, network design, logistics, or sports scheduling. The goal of integer optimization is to solve a system of constraints over many discrete variables, and to find solutions that are 'good' in terms of given optimization criteria. While fast general-purpose algorithms for solving large systems of linear inequalities over continuous variables are well-established (linear programming), integer optimization problems which include discrete decisions pose a difficult challenge to algorithmics. Yet, discrete decisions ("the truck leaves *either* today *or* tomorrow") are a critical part of most real-world planning and scheduling scenarios.

In the recent past, the field of integer and combinatorial optimization has gained momentum, and amongst the many new algorithms, *heuristics* have taken a leading role in finding near-optimal solutions to specific optimization tasks. The success of special-purpose heuristics is mainly due to their effectiveness for large practical problems – even if they come with no theoretical guarantee of optimality. The drawback of special-purpose algorithms, however, is their limited applicability. As a result, many practical optimization problems are still attacked in an ad-hoc fashion since practitioners often lack the time and expertise to research and develop effective special-purpose algorithms for the diverse optimization problems that arise.

This monograph explores a new domain-independent approach to integer optimization, which, unlike traditional strategies for integer optimization, is based on local search. It develops the central ideas and strategies of *integer local search* and describes possible combinations with classical methods, such as linear programming. In a number of case studies, it demonstrates the surprising effectiveness of the approach for a variety of realistic discrete optimization problems.

Like traditional strategies for integer linear programming, integer local search operates on an abstract model of the problem to be solved and can thereby exploit the underlying commonalities shared by many real-world problems. As a result, solvers based on the technology described here can be combined with existing off-the-shelf modeling languages for integer programming and can be applied to many integer optimization problems without the need of code implementation. We investigate the potential of integer local search for various domains (time tabling, sports scheduling, radar surveillance, course assignment, and capacitated production planning) and compare the experimental results to state-of-the-art integer programming and constraint programming approaches.

This book is written for researchers and practitioners in the area of combinatorial optimization from artificial intelligence and operations research. Developers with an interest in the design of optimization algorithms will benefit from the detailed description of new local search strategies. Practitioners in the field can obtain insights into modeling issues and learn about the capabilities of integer local search, which often surpasses state-of-the-art IP solvers for the domains under investigation.

This book is organized as follows. Chapter 1 introduces the context in which this work is situated, integer optimization, heuristics, and local search. It provides a high-level description of the basic strategy of integer local search and the underlying representation of over-constrained integer programs. It also outlines the experimental results from the application case studies. Chapter 2 briefly introduces important general frameworks for combinatorial optimization and their terminology, i. e. integer linear programming branch-and-bound, finite domain constraint programming, and local search. It also discusses complementary search relaxations as a new characteristic to classify optimization methods.

Chapter 3 contains the technical contributions, over-constrained integer programs and presents an in-depth description of the integer local search method WSAT(OIP). It also discusses several possible combinations with linear programming and illustrates different variations of the WSAT(OIP) strategy with graphical examples. Chapter 4 describes the case study methodology. It discusses criteria of success for practical optimization methods and motivates the experimental design and the problem selection.

The remainder of the text describes the case studies, each chapter focusing on a particular problem type, and providing evidence that important criteria of success are met by integer local search. Chapter 5 describes applications in time-tabling and sports scheduling, Chapter 6 radar surveillance and course assignment, and Chapter 7 presents an application to capacitated production planning. Chapter 8 finally discusses limitations and extensions of the current methods and concludes with suggestions for future work.

Acknowledgements

I am indebted to Gert Smolka for his support, guidance, and advice during the course of my doctoral research, which he supervised and on which this book is based. Thanks also for encouraging me to publish this monograph.

I am grateful to Henry Kautz for the insipring work of his group at AT&T Labs, which provided the starting point for this work. It is no exaggeration to say that without his and Bart Selman's work on local search, this book wouldn't exist. Many thanks to Henry also for co-examining my thesis and for much support during a visit at AT&T Shannon Labs.

Further, I thank Jimi Crawford for his support and encouragement over the years. Jimi has had a profound influence on my research orientation since introducing me to local search at CIRL in 1995. He invited me to i2 Technologies and made it possible to connect this research to manufacturing planning problems.

Several colleagues have contributed to this research through important discussions, particularly Martin Henz, Martin Müller, and Andrew Parkes. Seif Haridi and Per Brand provided the right application problem at the right time and thereby sparked initial ideas for this work. I have benefitted from discussions with Joachim Niehren, Christian Schulte, Alexander Bockmayr, Thomas Kasper, Mukesh Dalal, Ramesh Iyer, Narayan Venkatasubramanyan, Jörg Würtz, David McAllester, and Mats Carlsson. Thanks to Ramesh and Narayan for generously allowing me to include material from a joint publication, and to i2 and a client for making the publication possible. Martin Henz, Martin Müller, Michael Trick, Tobias Müller, and Joachim Niehren commented on draft chapters. Thanks to Michael Trick and George Nemhauser for sharing the ACC problem requirements, to David Abramson and Marcus Randall for providing the GPSIMAN solver of David Connolly, and to David Gay for tips on AMPL. I have enjoyed working with all the members of the Programming Systems Lab, who introduced me to the fascinating world of Oz. Special thanks to Ralf Scheidhauer and Michael Mehl for system-related help.

I am deeply grateful to my parents, Irene and Peter, for their love and support in good times and bad. And to Christine, for her wise advice and for all the love we share.

This research was supported by a doctoral scholarship of the Deutsche Forschungsgemeinschaft (DFG) within the 'Graduiertenkolleg Kognitionswissenschaft', Saarbrücken, Germany. The work was carried out in the computer science department of the Universität des Saarlandes between November 1995 and August 1998, and during research stays in the optimization team of i2 Technologies in Summer 1997, at AT&T Shannon Labs, the National University of Singapore and SICS in 1998, and at DFKI Saarbrücken in 1999.

February 1999

JOACHIM P. WALSER
Saarbrücken, Germany

Contents

List of Figures

List of Tables

1. Introduction

"Integer programming has gone through many phases in the last three decades, spurred by the recognition that its domain encompasses a wide range of important and challenging practical applications."

Fred Glover in [56], 1986

Integer and combinatorial optimization problems arise when a large number of discrete organizational decisions have to be made, subject to constraints and optimization criteria. This monograph describes and investigates new domain-independent local search strategies for linear integer optimization. This chapter briefly introduces integer optimization and heuristics, and presents an outline of *integer local search*, the approach to integer optimization that is the subject of this book. Integer local search generalizes local search for propositional satisfiability to linear integer optimization. Research in this area is situated in the interface between artificial intelligence and operations research. Since the two fields have been relatively separated in their past, terminology conflicts occasionally arise that we will attempt to point out.

1.1 Integer Optimization and Heuristics

A major challenge in algorithmics to date is to devise efficient and robust methods for combinatorial optimization problems. Combinatorial optimization is the problem of solving a system of constraints over many discrete variables and finding solutions that maximize or minimize some optimization criteria [117, 114]. The complexity of many interesting combinatorial optimization problems is known to be *NP*-hard.[1]

Combinatorial optimization problems vary largely, prototypical examples ranging from time-tabling over machine-scheduling to resource allocation. A problem class that captures a wide range of practically important problems

---

[1] A *search* problem X is *NP*-hard if for some *NP*-complete *decision* problem Y there is a polynomial-time reduction from Y to X [87].

is the *integer linear programming problem* (ILP).[2] An ILP consists of a set of linear inequalities (constraints) over integer variables, and a linear *objective function*, and is usually defined [114] as:

$$\text{(ILP)} \qquad \min \{ \, \mathbf{cx} \, : \, Ax \geq \mathbf{b}, \, \mathbf{x} \in \mathbf{Z}_+^n \, \},$$

where $\mathbf{Z}_+^n$ is the set of nonnegative integral n-dimensional vectors and $\mathbf{x} = (x_1, \ldots, x_n)$ are the variables. An *instance* of the problem is specified by the *data* $(\mathbf{c}, A, \mathbf{b})$, with $\mathbf{b}$ and $\mathbf{c}$ n-vectors and A an $m \times n$ matrix, and all numbers are rational (note that an equality constraint can be represented by two inequalities). If all variables are binary (0-1), the problem is also called 0-1 ILP. Throughout this book, we assume the optimization objective to be minimizing, and focus on integer programming problems with *linear* constraints and objective functions. A variable assignment that meets all constraints is called a *feasible solution*.

A wide variety of methods have been developed for solving ILP problems and are the subject of ongoing research in mathematical programming. When the integrality restrictions ($\mathbf{x} \in \mathbf{Z}_+^n$) of an ILP are relaxed, one obtains the well-studied linear programming problem (LP) for which polynomial algorithms are known [92] and efficient implementations exist. ILP methods can make use of LP relaxations in many ways, for instance for lower bounding or feasibility testing, and most integer programming (IP) frameworks are based on iteratively solving LP relaxations, e. g. branch-and-bound or cutting-plane algorithms [114]. In summary, ILP provides a good starting point as representation for general-purpose optimization methods.

Domain-Specific vs. Domain-Independent Techniques.

There are two orientations of research on optimization algorithms. First, specialized techniques that excel in solving narrow classes of optimization problems for which maximal quality is crucial and development times can be neglected. Such techniques often solve sub-problems of the general ILP problem, for example the *Set-Covering Problem* (SCP).

Let $M = \{1, \ldots, m\}$ be a finite set and let $\{M_j\}$ for $j \in I = \{1, \ldots, n\}$ be a given collection of subsets of M. We say that $F \subseteq I$ is a *cover* of M if $\bigcup_{j \in F} M_j = M$. In the set-covering problem, c_j is a cost associated with M_j, and we seek a minimum-cost cover [114]. SCP can be formulated as an integer linear program using 0-1 variables x_j with $x_j = 1$ if and only if j is in the cover.[3] The ILP formulation of SCP is:

$$\text{(SCP)} \quad \min \{ \, \mathbf{cx} \, : \, \sum_{j=1}^{n} a_{ij} x_j \geq 1, \, i = 1, \ldots, m, \; \mathbf{x} \in \{0,1\}^n \, \},$$

---

[2] The term 'programming' dates back to the 1940s, when Dantzig described the simplex method for linear programming. 'Programming' was a military term that, at the time, referred to planning and scheduling of logistics.

[3] $x_j \in \{0,1\}$ can be constrained through inequalities $x_j \geq 0, -x_j \geq -1$.

where c is an n-vector, (a_{ij}) is a 0-1 matrix, and the variables are binary (0-1). Domain-specific strategies for the set-covering problem are restricted to problems of the form (SCP) as specified. Incorporating other constraints typically requires adjusting the algorithms or replacing the strategies altogether.

In contrast, domain-independent techniques strive to be flexible and applicable to a wider range of practical problems without the need of designing strategies on a class-by-class basis [59]. Such techniques work from a *model* of a given problem instance (a representation in a suitable constraint class). They are of practical importance because practitioners often lack the necessary time and expertise to research and develop effective special-purpose algorithms. Further, in real environments, flexibility is often critical to respond to rapidly changing requirements.

Research on domain-independent techniques for combinatorial optimization has given rise to general-purpose tools in integer programming, such as a variety of branch-and-bound solvers (e. g. CPLEX, LINDO, XPRESSMP, MINTO to name but a few). More recently, Constraint Programming (CP) systems have entered the picture that support rapid development of domain-specific methods and incorporate an increasing variety of techniques for constraint propagation and search (e. g. CHIP [42], Oz [137], ILOG solver [120]).

Heuristics and Local Search. Optimization methods can either be *exact* or *approximate*. While exact methods perform a systematic search for optimal solutions, approximate methods provide no theoretical guarantee for finding optimal or even feasible solutions. In operations research, approximate methods are commonly termed *heuristics*, and we will stay with this usage.[4] Heuristics concentrate on finding near-optimal solutions quickly, and have received much interest in recent years due to their practical success [126, 123, 1, 61].

An important class of heuristics is *local search* [1] which has a long history for combinatorial optimization and dates back to methods for the traveling salesman problem in the 1950s and 1960s [20, 37, 102]. The key idea behind local search is to start from a solution and iteratively perform changes to improve it. There are many variations of local search methods, sharing the common notion of *local moves* which are transitions in the space of (feasible and possibly infeasible) solutions, typically according to a strategy that works by improving the *local gradient* of a measure of the solution quality (a strategy called *hillclimbing*). Many variants of local search exist that can be applied to combinatorial optimization problems, prominent examples being simulated annealing [96] tabu search [57, 61], genetic algorithms [113, 62] or the greedy randomized adaptive search procedure (GRASP) [128].

In artificial intelligence, local search strategies have recently seen much success for model finding in propositional satisfiability [135, 64, 111, 53, 134]

---

[4] Note that in artificial intelligence, the term 'heuristic' commonly refers to a 'rule-of-thumb' decision strategy of an algorithm.

and a variety of applications to combinatorial problems have been reported [132, 35, 128, 54, 93]. Local search strategies of this kind are also called *iterative repair* [108, 155, 135]: Given a problem that is stated in terms of variables and constraints, one first generates some initial assignment of all variables, normally violating a number of constraints. Subsequently, variable values are changed in order to reduce the number of conflicts with the constraints, i.e. in order to *repair* the current variable assignment.

Heuristics for Integer Optimization. Most heuristics for integer optimization are dedicated to a specific problem (like set-covering or job-shop scheduling) and often excel in terms of the quality of solutions found and efficiency. Perhaps surprisingly, only few efforts have been made to devise heuristics that target a wider range of combinatorial optimization problems and operate on problem representations using constraints.

Recently, several general-purpose heuristics have been described which aim at solving general ILP problems (some being extensions of the pioneering work by Balas and Martin [9]). These heuristics are of two types, (i) approaches which relax the integrality constraints and primarily operate on continuous variables [9, 3, 104, 59, 61] (e.g. by solving the linear program followed by special pivot moves), and (ii) local search methods in which the local moves are performed directly in the space of integer solutions, such as simulated annealing [33, 4] and stochastic local search [149, 150].[5]

The methods presented in this monograph are of the second type and arise from generalizing successful strategies of local search for propositional satisfiability [149, 150]. We will refer to these methods as *integer local search*.

Local Search for Propositional Satisfiability. A problem of much interest in computer science is the *propositional satisfiability problem* (SAT). Let V be a finite set of 0-1 (*Boolean*) variables. An assignment for V is a mapping from V to $\{0, 1\}$. A *literal* is either a variable v or its negation $\bar{v}$ ($\bar{v} = 1$ iff $v = 0$). A *clause* is a set of literals, and is *satisfied* by a given assignment A if at least one of its literals is assigned to 1. A set of clauses C is interpreted in *conjunctive normal form* (CNF): An assignment A satisfies C if and only if all clauses in C are satisfied under A.

(SAT)   Given a set V of variables and a set C of clauses over V, is there an assignment for V that satisfies all clauses in C?

We observe that SAT is also a special case of 0-1 ILP (with objective function $c = 0$), since each clause can be translated to a linear inequality: For instance, the clause $\{\bar{x}, y\}$ can be translated to $(1 - x) + y \geq 1$. The 0-1 ILP problem is more general than SAT because it allows for arbitrary right-hand-sides (**b**) and coefficients (**A**), and because SAT is a *decision problem*, i.e. there is no explicit representation of an objective function in SAT.

---

[5] [9, 3, 104] actually contain *phases* of type (ii).

The Walksat Strategy. A number of efficient local search strategies have been developed for SAT in recent years, one of the most successful ones being the Walksat procedure by Selman, Kautz, and Cohen [134, 106]. To find solutions to a set $\mathcal{C}$, the basic Walksat strategy performs a greedy local search equipped with a 'noise' strategy: Initially, all variables are assigned a random value from $\{0, 1\}$. While some of the clauses will be satisfied, others are violated. To seek an assignment that satisfies all clauses, the method iteratively selects a violated clause $c \in \mathcal{C}$, and from c selects a variable such that changing its value yields the largest increase in the total number of satisfied clauses. If no variable exists that improves the total number of satisfied clauses, a variable from c is selected at random according to some detailed scheme. Such variable changes (*flips*) are repeated a fixed maximal number of iterations after which a restart takes place. If no satisfying assignment is found after a fixed number of restarts, the procedure is terminated unsuccessfully.

1.2 Integer Local Search

This monograph is concerned with local search strategies for integer optimization. It describes, discusses and empirically analyzes WSAT(OIP), a domain-independent method that generalizes local search for propositional satisfiability (the Walksat strategy) to integer optimization and integer constraint solving.

Over-constrained Integer Programs. Many integer optimization problems have no concise SAT encoding, and hence SAT local search algorithms cannot be applied. In order to generalize SAT local search to integer optimization, we introduce an extension of SAT to a constraint system called *over-constrained integer programs* (OIPs). Extending the repair strategy of Walksat to OIP optimization will then be a natural step.

An OIP consists of hard and soft inequality constraints, wherein the optimization objectives are represented by the soft constraints. If all inequalities are linear, the OIP problem can be formulated in matrix notation as

$$Ax \geq b, \quad Cx \leq d \ (soft), \quad x \in D,$$

where A and C are $m \times n$-matrices, $\mathbf{b}, \mathbf{d}$ are m-vectors, and $\mathbf{x} = (x_1, \ldots, x_n)$ is the variable vector, ranging over positive finite domains $x_i \in D_i$. A variable assignment that satisfies all hard constraints is called a *feasible solution*. Given a tuple $(A, \mathbf{b}, C, \mathbf{d}, \mathbf{D})$, the OIP minimization problem is

$$(\text{OIP}) \quad \min \{ \ \|C\mathbf{x} - \mathbf{d}\| \ : \ Ax \geq b, \ \mathbf{x} \in D \ \}, \quad \|\mathbf{v}\| := \sum_i \max(0, v_i),$$

wherein the objective is to find a feasible solution with minimal soft constraint violation. In $\|.\|$, the contribution of each violated soft constraint to the overall objective is its degree of violation. As will be shown, OIPs are a special case of integer linear programs because each soft constraint encodes a piecewise-linear convex objective function. The reduction enables effective combinations with linear programming for lower bounding, initialization by rounding, search space reduction, and feasibility testing.

While ILP encodes the optimization objectives using a monolithic objective function, OIP uses many competing soft-constraints. OIPs are a natural representation to generalize iterative repair strategies like Walksat: Only violated constraints need to be repaired, and no principal distinction is drawn between repairing hard and soft constraints, thereby seeking solutions that are both feasible and near-optimal. OIP is similar enough to ILP to make use of algebraic modeling languages like AMPL [48] as front-end to an integer local search solver.

Unlike methods that rely on properties of the linear relaxation, note that integer local search is not limited to inequality and equation constraints but can be extended to other types of constraints, for example disequality constraints ($x \neq y$) or symbolic constraints ($all\text{-}different(x_1, \ldots, x_n)$).

The WSAT(OIP) **Strategy.** WSAT(OIP) is now a natural generalization of Walksat. It performs local moves in the space of feasible and infeasible solutions by repairing constraint violations. Iteratively, WSAT(OIP) changes a variable value as follows. First, select a violated constraint c: if only hard or only soft constraints are violated, select c at random. If hard *and* soft constraints are violated, with some probability p_{hard} select a random violated hard constraint and with $1 - p_{hard}$ a random violated soft constraint. The p_{hard} parameter controls how quickly the search is driven into the feasible region of the search space.

Next, to decide which variable from c to change, it no longer suffices to reduce the number of violated constraints as Walksat does. Instead, a *score* is maintained that accounts for both the soft constraint violation and the degree of *infeasibility* of an assignment (using $\|.\|$ above for both soft and hard constraints). For 0-1 variables, a variable change amounts to flipping the value. To extend the principle to finite domain variables, a new class of local moves is introduced that changes a variable value to a smaller or greater values nearby. Like variable flips, such *trigger* moves are also induced by violated linear inequalities and only repairs in c are considered that reduce the violation of c.

Among the possible repairs, the strategy greedily chooses the one that most improves the overall score, but occasionally resorts to random moves if no improvement is possible. Following Walksat, WSAT(OIP) is stochastic local search and inherits Walksat's noise strategy. To improve the performance for realistic problems, the method is further extended with principles from tabu search [57, 58] by keeping a history of past moves. History information can be used to diversify the search process. For instance, variables can have a tabu status that disallows changes that would undo a recent change (tabu status can be overridden by *aspiration criteria*). Finally, detailed tie breaking rules [53] are built into the strategy that are based on the history of the search and decide which move to perform when several variables have an equal score.

1.3 Experimental Results

A substantial part of this book is concerned with an empirical investigation of the WSAT(OIP) method for practical optimization problems. The problems under consideration are a resource allocation problem termed 'the Progressive Party Problem' (PPP) [136], the scheduling of a basketball season (ACC) [115], radar surveillance (RS) [24], course assignment (CA), and capacitated production planning (CPP) [150]. The problem instances stem from real applications (ACC, CA, CPP) or contain realistic structure (PPP, RS). Each class contains large-scale instances, involving several thousand variables and constraints.

We compare the results of integer local search WSAT(OIP) to approaches from the recent literature (ACC [115, 69], PPP [136, 76]), and to some of the best state-of-the-art methods for the respective problems. In particular, we compare the results to IP branch-and-bound (CPLEX 5.0 [80], an efficient general-purpose mixed integer programming solver), several constraint programming approaches and a previous general-purpose simulated annealing method for 0-1 ILPs (GPSIMAN) [33].

The problems fall into three broad classes, grouped in three experimental chapters: (i) tightly constrained 0-1 integer constraint problems (PPP, ACC), (ii) 0-1 integer optimization problems (RS, CA), and (iii) integer optimization problems with finite domain variables (CPP).

Because the problems vary remarkably, the experimental results demonstrate the general-purpose nature of the method. Further, the different types of problems will shed light on different performance aspects: Scaling of runtime with increasing size, scaling with increasing constrainedness, and residual robustness (performance variation on a distribution of similar instances). The experimental results can be summarized according to the three problem types.

(i) *0-1 integer constraint problems* from time-tabling (PPP, ACC). Even though such feasibility problems can be formulated as ILPs, these encodings pose a difficult challenge for integer programming strategies [136, 76, 115, 149]. Our case studies demonstrate that integer local search can solve such problems directly from a 0-1 representation.

For PPP, a constraint programming approach is known [136] to solve individual instances of the problem, while the problem appears to be beyond the size limitations of integer programming branch-and-bound [136, 76]. Our experiments show that the core problem can be solved more efficiently using WSAT(OIP). Further, when slight variations of the instance given in [136] are considered, we find that local search is robust with respect to the modifications, while we were not able to find a CP enumeration strategy to solve all test problems.

The second problem, ACC basketball scheduling, was presented and solved in a multi-stage approach in [115]. The approach uses IP branch-and-bound and explicit enumeration in four separate stages, with a performance of roughly 24h to generate a set of feasible schedules. A more efficient CP strategy is also known [69]. We present a 0-1 linear integer encoding of the problem that does not separate the problem into stages. Given this monolithic 0-1 representation, we study the performance of integer local search with increasing problem constrainedness. Our experiments show that for loosely constrained instances, integer local search finds solutions in seconds. As the constraints are tightened, runtimes tend to increase moderately. When the entire constraints from [115, 141] are encoded, we show that WSAT(OIP) still finds solutions in 30 minutes, despite only 87 solutions remain! In experiments with IP branch-and-bound we find that CPLEX manages to solve only the loosely constrained instances within reasonable runtime.

(ii) *0-1 integer optimization problems* (RS / CA — extended set covering / generalized assignment) are classical domains of IP branch-and-bound methods, and pose a difficult challenge for constraint programming approaches.

On these problems, the experiments show that WSAT(OIP) and CPLEX efficiently find solutions of similar quality for small problems. As the problem size increases, the runtime of WSAT(OIP) scales gracefully. Integer local search therefore outperforms IP branch-and-bound on large problems by orders of magnitude.

In one series of experiments on (RS), we further find that the OIP representation (soft constraints) is critical for WSAT(OIP) to achieve the above reported performance. The results show that the OIP representation captures structural aspects that can be exploited by integer local search, leading to a promising primal heuristic for large-scale problems.

(iii) *OIPs with finite domain variables.* The last experimental chapter studies a real production planning problem (CPP) from the process industry (manufacturing of chemicals, food, plastics, etc.), technically termed 'capacitated lot sizing'.

Because the problem requires discrete lot-sizes, domain-specific methods from the literature are not directly applicable. We therefore approach the problem with WSAT(OIP), and empirically compare the results to a mixed integer programming branch-and-bound approach (again using CPLEX) on real problem data. We find that integer local search is considerably more robust than MIP branch-and-bound in finding feasible solutions in limited time, in particular as the capacity constraints are tightened.

With respect to production cost, both methods find solutions of similar quality.

With respect to parameters for WSAT(OIP) we find that a set of standard parameters usually performs well on the given domains without much manual fine-tuning, and little further customization was necessary, with exception of constraint-weights (used for CPP), and redundant constraints (used for ACC).

GPSIMAN. Where applicable, we additionally perform experiments with GP-SIMAN, a general-purpose simulated annealing method [33]. GPSIMAN was not able to find solutions to the 0-1 constraint problems, even at the lowest tightness levels within 12 hours. For the optimization problems under consideration, GPSIMAN was able to solve instances of only one class (RS), but did not succeed in solving the larger instances satisfactorily.

Limitations and Scope. The computational study of integer local search described here contains several limitations. First, throughout our experiments, we assume *pure* integer optimization problems. Industrial problems, however, often contain a continuous component and require handling mixed integer programming problems, which are not addressed in the framework of integer local search as presented in this monograph.

Second, our case studies do not investigate problems that contain a strong intricate solution structure, like traveling salesman problems or job-shop scheduling. Encodings of these problems into ILP are typically large and the solution structure is difficult to maintain by local moves that change single variable-values, as performed by WSAT(OIP). For such problems, more informed local moves need to be considered. Another general and effective strategy for such problems is "Abstract Local Search" by Crawford, Dalal and Walser [36]. In Abstract Local Search, a greedy constructive heuristic is combined with local moves in an abstract search space.

Third, with exception to one problem class (CPP), we observe that the instances under consideration contain mainly 0-1 *coefficients*. In fact, for the basic version of WSAT(OIP), strongly differing coefficients are problematic because the score gradient favors variables that appear with large coefficients. An extension of the basic scoring scheme will be discussed that addresses larger coefficients.

1.4 Research Contributions

Although some earlier research papers have described basic strategies and case studies of WSAT(OIP) [149, 150] and several related general-purpose heuristics have been reported [33, 116, 4, 5], this monograph represents the first in-depth treatment of integer local search. It advances the state-of-the-art of heuristics for integer optimization by the following contributions.

Generalization of SAT local search to linear integer optimization over binary (0-1) and finite domain variables.

Development of over-constrained integer programs (OIP) as a basis for applying iterative repair to integer optimization. Reduction from OIP to ILP in order to permit combinations with linear programming.

Engineering of an effective local search strategy for over-constrained integer programs (WSAT(OIP)) including local moves for 'triggering' finite-domain variable values.

Description of strategy enhancements used by an efficient solver, WSAT(OIP), which interfaces with AMPL and CPLEX, and which provides the first local search solver for integer optimization to interface with an algebraic modeling language.

Development of the notion of integer local search and its characterization in terms of search relaxations.

Empirical study of the approach demonstrating its flexibility, efficiency and graceful scaling for a variety of hard realistic integer optimization / integer constraint problems. Within the empirical study, the original results include
   – evidence that large or tightly constrained problems beyond the limitations of ILP branch-and-bound can be solved directly from a monolithic 0-1 integer encoding by integer local search,
   – evidence that OIP soft-constraint encodings can capture structural aspects exploitable by integer local search,
   – two problem-specific results: (i) a heuristic can solve a temporally dense double-round robin sports scheduling problem, and (ii) a general-purpose heuristic can solve a large production planning problem encoded with finite domain variables.

2. Frameworks for Combinatorial Optimization

This chapter outlines some successful frameworks for combinatorial optimization. It is meant as a brief introduction into terminology and outlines the basic principles of the frameworks which are applied in the subsequent case studies. The three frameworks that will be discussed, integer linear programming (ILP), finite domain constraint programming (CP) and local search are well established, comprise of a variety of techniques, and many successful applications have been reported. ILP and CP can be considered the state-of-the-art of general-purpose optimization methods, whereas local search should be seen as an approach that can be tailored to many different optimization problems by adapting its conceptual components to the respective problem context. We also discuss successful local search methods for solving propositional satisfiability problems.

2.1 Integer Programming Branch-and-Bound

One of the most successful and widespread approaches to general integer programming problems is the *IP branch-and-bound* algorithm. Here, we will discuss the branch-and-bound algorithm that builds on linear programming relaxations. It is the basic algorithm used by most commercial codes for solving *mixed-integer programming problems* (MILP). MILP is a superset of pure integer programming ILP, in that some of the variables are required to be integer, yet others may assume continuous values. The following short exposition follows Nemhauser & Wolsey [114]; for a more complete introduction and for an introduction to the underlying strategy of *linear programming* (LP), the reader is referred to an introductory text in operations research [114, 71, 31].

IP branch-and-bound is concerned with finding optimal solutions to the ILP problem,

$$(\text{ILP}) \qquad z_{\text{IP}} = \min\{ \mathbf{cx} \; : \; \mathbf{x} \in S \; \}, \text{ where } S = \{\mathbf{x} \in \mathbf{Z}_+^n \; : \; A\mathbf{x} \geq \mathbf{b} \; \},$$

Note that the ILP becomes a MILP problem when some of the integrality constraints are relaxed. The branch-and-bound strategy can equally be applied to MILP problems. The basic principle underlying branch-and-bound is to *divide-and-conquer*. The original problem is divided into sub-problems, by partitioning the set of feasible solutions into subsets, and recursively solving those. This *branching* strategy continues until the subproblem is either small enough to be solved directly or it is known that it can be discarded because it cannot possibly contain an optimal solution for the original problem (*bounding*).

LP relaxation. In IP branch-and-bound, a tree is searched. Initially, at the root node, a *linear relaxation* of the original problem is solved, i.e. S is replaced with $S_{\mathrm{LP}}^0 = \{\, x \in \mathbf{R}_+^n \ : \ Ax \geq b \,\}$. Subsequently, constraints are added to the problem to divide it into subproblems to be solved separately. At each node, a linear relaxation is solved,

$$ z_{\mathrm{LP}}^i = \min\{\, \mathbf{c}\mathbf{x} \ : \ \mathbf{x} \in S_{\mathrm{LP}}^i \,\}, $$

where S_{LP}^i constitutes the linear relaxation of the respective subproblem S^i.

Also, at any node i during search, there are three conditions when a node can be *pruned*, i.e. none of its children need to be considered. The conditions are

(i) *infeasibility*, if $S_{\mathrm{LP}}^i = \emptyset$, i.e. no solution to the linear program exists
(ii) *optimality*, if $\mathbf{x}^i \in \mathbf{Z}_+^n$, i.e. the solution is integral
(iii) *value dominance*, $z_{\mathrm{LP}}^i \geq \overline{z_{\mathrm{IP}}}$, where $\overline{z_{\mathrm{IP}}}$ is the value of a known feasible solution to ILP.

Tree Search. If the node is not pruned by one of the conditions (i)-(iii), branching occurs. Branching is done by adding linear constraints to the current problem. There are various ways to perform branching. One common way is to perform 'variable dichotomy', that is choose some variable x_j which has a non-integral solution value x_j^* in the linear relaxation. Branch on $S \cap \{\mathbf{x} \in \mathbf{Z}_+^n \ : \ x_j \leq \lfloor x_j^* \rfloor\}$ and $S \cap \{\mathbf{x} \in \mathbf{Z}_+^n \ : \ x_j \geq \lfloor x_j^* \rfloor + 1\}$, and recursively search both nodes. At any node i, if the given LP solution is optimal, the current feasible solution is stored.

The size of the enumeration tree depends largely on the pruning, i.e. on the quality of the bounds produced by the linear programming relaxation and on the solutions found early in the search. If the linear relaxation at the root yields solutions whose objective function values are close to the optimal solutions of (ILP), we say that the relaxation is *tight*.

Many issues need to be considered to develop efficient branch-and-bound strategies, such as the selection of branching variables, and more generally search strategies to explore the search tree. Efficient branch-and-bound implementations like CPLEX further add valid inequalities (also called cuts) which are inferred from specific classes of constraints present in the original constraints. Using these components in addition to efficient algorithms for computing and re-computing LP relaxations, ILP branch-and-bound provides a general and efficient technique for many ILP optimization problems. A variety of efficient commercial branch-and-bound solvers are available on the market, (e.g. CPLEX, LINDO, XPRESSMP, MINTO).

2.2 Finite Domain Constraint Programming

Finite domain constraint programming (CP) is a programming technique designed for solving *constraint satisfaction problems* (CSP). A CSP is typically defined [142] in terms of (i) a set of variables, each ranging over a finite discrete domain of values, and (ii) a set of constraints, which are relations over subsets of the variable domains. The problem is to assign values to all variables from their domains, subject to the constraints. A large number of problems in AI and other areas of computer science can be viewed as special cases of this general notion of CSP [98].

Constraint programming evolved from research in constraint logic programming languages ([82] gives a survey), and led to the development of constraint programming languages such as CHIP [42], Oz [137], CLP(FD) [32], ECLiPSe [6], and the constraint programming library ILOG Solver [79, 120]. *Concurrent* constraint programming properly addresses the concurrent aspects of constraint programming [129, 70, 137]. CP technology is becoming increasingly successful for industrial problems, and many emerging applications have been attacked with constraint programming (see [146] for a survey). We will next describe the CP framework, following [69, 139].

Constraint Store. Every variable of the CSP problem is represented by a finite domain variable. All information on the variables is stored directly in a *constraint store* [81], in terms of the sets of possible values that each variables can take. This set is called the current domain of the variable. Computation starts with the current domain of every variable being the one given in the CSP. For example $x \in \{1, \dots, 5\}$ might be the initial current domain. Subsequently, the information on the current variable domains is updated in the constraint store. This means that the domain information of the variables is treated as a *primitive* constraint.

Constraint Propagators. In addition to the primitive constraints, combinatorial problems require *non-primitive* constraints, which are handled through computational agents called *propagators* [153, 138] as illustrated in Figure 2.1. Propagators operationalize the semantic definition of the constraints. Some of the corresponding local consistency techniques have been established in artificial intelligence [142]. In finite domain constraint programming, the basic inference principle of propagators is *domain-reduction*. Each propagator observes the variables that occur in the constraint. Whenever possible, it amplifies the constraint store in terms of these variables by excluding values from their domains.

For example, a propagator can realize a constraint $all\text{-}different(x_1, \dots, x_n)$, requiring that all variables $x_1, \dots, x_n$ be assigned different values. There are different ways to operationalize this constraint, some based on sophisticated algorithms [127, 121]. A straightforward way is to wait until the domain of one of the variables becomes a singleton and then eliminate its element from the domains of the other variables. Removing the value from the domains of

Propagator   ...   Propagator

Constraint Store

Figure 2.1. Computational setup of a computation space.

the other variables may in turn trigger other propagation. If a propagator is entailed, it ceases existence. If a propagator can determine that a variable domain becomes empty, the constraint store becomes inconsistent.

The propagation process continues until no propagator can further amplify the constraint store. The *computation space* (i. e. the constraint store together with the propagators) is said to become stable. Still, at this point, many variables will typically have multiple values in their domains. Thus the constraint store does not represent a solution to the CSP.

Tree Search. Once a constraint store has become stable, no further information can be inferred by the propagators. To solve the current finite domain problem P^i at node i, one can choose a constraint C and solve two subproblems $P^i \cup \{C\}$ and $P^i \cup \{\neg C\}$. Such *branching* on constraint C is likely to trigger new propagation. After stability is reached again, the branching process is continued recursively on both sides. Often, the constraint C is an assignment of a variable to one of its remaining values, e. g. $x = 1$ if $\{1, \ldots, 3\}$ are remaining.

To obtain a complete search for all solutions, the branching continues recursively until either all variables have a unique value in the store or until the store has become inconsistent. The branching scheme yields a complete set of solutions to the original problem because of the soundness and completeness of the branching. The choice of constraint C is a critical issue for the performance of the search. If C assigns one variable, an *enumeration strategy* determines which variable and which value from its domain are selected for assignment. The issue here is to find good heuristics for variable and value selection. A full description of the branching is called a search strategy.

Finite domain CP systems usually offer rich support for the propagators that can be used to model constraint problems. Also CP systems take control of propagation and search. Different CP systems offer varying support in the design of search strategies. An advanced concept for high-level programming of search strategies is through using first-class computation spaces [131]. By providing the building blocks of constraint propagation, branching and search, constraint programming can be viewed as a supporting framework for the rapid development of domain-specific optimization strategies.

CP has seen much success in a variety of domains; for instance in scheduling, various algorithmic techniques from OR (e. g. [26, 8]) have been integrated into constraint programming by encapsulation into propagators and branching strategies [10, 153]. Thos strategies are available in both academic and commercial solvers such as Oz or CHIP, or ILOG Scheduler.

2.3 Local Search

Many combinatorial optimization problems are *NP*-hard [50], and the theory of *NP*-completeness has reduced hopes that *NP*-hard problems can be solved within polynomially bounded computation times. Nevertheless, sub-optimal solutions are sometimes easy to find. Consequently, there is much interest in approximation algorithms (*heuristics*) that can find near-optimal solutions within reasonable running times, even at the cost of giving up the optimality guarantee under infinite runtime.

Practically, for many realistic optimization problems good solutions can be found efficiently, and heuristics are typically among the best strategies in terms of efficiency and solution quality for problems of realistic size and complexity [1, 123, 61, 126]. Heuristics can be classified as either *constructive* (greedy) heuristics or as *local search* heuristics. The former are typically efficient one-pass algorithms whereas the latter are strategies of *iterative improvement*. We will be concerned exclusively with local search heuristics here. By relaxing the guarantee of finding optimal (or even feasible) solutions, heuristics are not required to search for solutions systematically. Instead they can freely follow a local gradient in the search space. It is sometimes argued that it is this freedom that gives local search strategies an advantage over systematic strategies on certain large optimization problems.

Local search algorithms have been applied to many combinatorial optimization problems in one or another variation. Local search algorithms can be described in terms of several basic components. The *combinatorial problem* to solve, a *cost function* of a solution to the problem, a *neighborhood function* that defines the possible transitions in the search space. And finally, the *control strategy* according to which the local moves are performed. In the following definitions, we will informally follow [1] extended to apply to feasibility problems as well.

Combinatorial Problem. For the purpose of this discussion, a *combinatorial problem* can be specified by a set of *problem instances* to be solved. A problem is either a minimization or maximization problem.

Cost Function. An instance of a combinatorial problem is defined by the set of *feasible solutions* and a *cost function* that maps each solution to a scalar cost. The problem is to find a globally optimal feasible solution, i. e. a feasible solution that optimizes the cost function. We will only consider minimization problems here.

Neighborhood Function. Local search progresses by making transitions from one *node* to another. The set of nodes includes the feasible solution, but may also include other, infeasible, solutions. Given an instance of a combinatorial problem, the *neighborhood function* is defined by a mapping from the set of nodes to its *neighbors*, i. e. the subsets of the set of nodes. A solution is *locally optimal* with respect to a neighborhood function if its cost is strictly better than the cost of each of all its neighbors. Notice that there can also be 'stateful' neighborhood functions, i.e. the set of neighbors can change as the search progresses.

Control Strategy. The final component is the control strategy. It defines the strategy of how the nodes are examined. For example, a basic control strategy of local search is *iterative improvement*. Here, one starts with some initial solution and searches its neighborhood for a solution of lower cost. If such a solution is found, the current solution is replaced and the search continues. Otherwise, the algorithm returns the current solution, which is then locally optimal.

A central problem of local search are *local optima*, i.e. nodes in the search space where no neighbor strictly improves over the current node in terms of the cost function. Many strategies have been proposed that address the problem of how to overcome local optima. In many cases, non-improving local moves are admitted based on a probabilistic decision (*noise*) or based on the history of the search. A number of 'meta-heuristics' have been proposed that address the problem of local optima.

2.3.1 Meta-heuristics

To obtain guiding principles for the design of effective optimization strategies, a number of conceptual meta-level strategies have been proposed for local search. These strategies are referred to as *meta-heuristics*, a term coined by Glover [56]. Some of the most prominent meta-heuristics for optimization simulate aspects of processes that occur in nature[1] (like genetic evolution, metal annealing or neural processes) and are briefly described in the following. Note that these strategies can typically be superimposed on a given basic control strategy.

Simulated Annealing, introduced for optimization by Kirkpatrick *et al.* in 1983, uses the metaphor of the annealing process in steal manufacturing by which the brittleness of the steal can be reduced [96]. The central idea is to accept a candidate move that decreases the solution quality based on a probabilistic decision. During the time of the search, the probability of acceptance of such deteriorating moves is decreased according to a given *annealing schedule* [2].

Genetic Algorithms model optimization as an evolutionary process. The strategy is to have a pool of chromosomes and iteratively apply the principles of mutation, mating and selection to attain 'survival of the fittest' [74, 62]. Genetic algorithms extend the basic local search scheme to *populations* of solutions. Early applications in Artificial Intelligence in the

---

[1] An exception is 'Tabu Search' framework.

1960s were in game-playing, pattern recognition, or adaptation, and it is more recently that applications to optimization problems have been reported [62, 125, 113].

Tabu Search uses information based on the history of the search [57, 61], and is a particularly successful strategy for many practical problems. The central idea is the use of adaptive memory to overcome local optima by driving the search to different (*diversification*) or back to promising (*intensification*) parts of the search space.

Artificial Neural Networks make use of the metaphor of the neuron and are organized in network structures. The network of nodes is iteratively modified by adjusting the interconnections between neurons according to various schemes. Neural networks have been popular since the late 1960s, and have more recently been successfully applied to optimization problems [78, 44, 119, 38].

Meta-heuristics express orthogonal concepts and hybrids are possible. Meta-heuristics are domain-independent conceptual strategies. For particular application problems, they require concrete implementation and their success varies between different application domains.

2.3.2 RISC and CISC Local Search

We can distinguish local search strategies for combinatorial optimization according to the structure of the local neighborhood. We have earlier given a short description of Walksat, a local search method for propositional satisfiability in which the local neighborhood of an assignment A consists of a subset of $\{A' : A' = A$ with one variable flipped$\}$ and contains $O(n)$ neighbors for a problem instance with n variables. This local neighborhood is simple and small.

On the other hand, many local search strategies use local neighborhoods that are more intricate and larger. For example, consider the *traveling salesman problem*, the problem of finding a minimal-cost *tour* visiting a number of cities such that each city is visited exactly once. Effective local search strategies for traveling salesman move in the space of feasible tours by complex edge-exchanges on tours [89]. Another example is job-shop scheduling, where the goal is to assign start-times to a number of tasks that compete for resources, such that the finishing time of the latest finishing task is minimized (makespan optimization). Effective techniques for job-shop-scheduling move in the space of feasible schedules by complex move operations between tasks that are on a *critical path*[2] of the schedule [7].

---

[2] A path through the directed graph of adjacent tasks which ends in one of the tasks that finish latest.

Such local search strategies differ from the SAT case in that problem solutions are expressed in terms of complex entities like *tours* or *schedules*. All solutions are required to comply with this structure, and to avoid the difficulty of restoring it, the local moves are designed to be structure-preserving. This is contrasted with local search for SAT and the proposed methods for integer local search, in which no explicit solution structure is preserved.

Here, we exclusively consider strategies which (i) operate by changing variable-values, (ii) have neighborhoods with a size linear in the problem size (number of variables), and (iii) guarantee to preserve only one property with any move, namely that all variables are assigned exactly one value from their domain. Because of this atomic nature of the local moves we can refer to strategies with properties (i)–(iii) as 'reduced instruction set' or *RISC local search*. This is in contrast with the *CISC local search* strategies for the domains traveling salesman or scheduling. We consider both feasibility and optimization problem formulations.

2.3.3 Local Search for SAT

Local search strategies have recently seen much success for model finding in propositional satisfiability [135, 64, 111, 53, 134]. Figure 2.2 gives the outline of a typical local search routine in the spirit of the now classic strategy GSAT by Selman, Levesque and Mitchel [135]. It searches for a satisfying variable assignment for a set of clauses A. Here, local moves are "flips" of variables that are chosen by *select-variable*, usually according to a randomized greedy strategy. The measure to perform hill-climbing on for the SAT problem is typically the number of satisfied clauses. The parameter Maxtries can be used to ensure termination, while Maxflips determines the frequency of restarts that can often help to overcome local minima (in the number of unsatisfied clauses).

Most SAT local search algorithms follow Figure 2.2 and thus have a remarkably simple structure on this level of abstraction. The remaining degree of freedom in this scheme is the policy to select the next variable to be flipped. This policy, in fact, is the key component of SAT local search algorithms and is of critical importance for their empirical performance. A variety of policies have been described [135, 64, 132, 134, 53, 118, 106] that vary in a static and a dynamic aspect. Statically, the probability of a variable to be flipped depends only on the current variable assignment and on the scheme to compute the *score* of a variable, e. g. how many clauses will be satisfied after the flip or how many clauses will break because of the flip.

Dynamically, the decision can also depend on the history of the search. For example, a Tabu element [61] can forbid to undo a recent flip. Alternatively, one can break ties between variables with an equal score according to how recently a variable has been flipped [53]

Randomness. The most effective local search procedures for SAT are randomized algorithms. There are two similar purposes for which randomness

```
proc Local-Search-SAT
      Input clauses C, Maxflips, and Maxtries
      Output a satisfying total assignment of C, if found
      for i := 1 to Maxtries do
            A := random truth assignment
            for j := 1 to Maxflips do
                  if A satisfies C then return A
                  P := select-variable(C, A)
                  A := A with P flipped
            end
      end
      return "No satisfying assignment found"
end
```

Figure 2.2. A generic local search procedure for SAT.

is used: (i) to select a starting point for the local search (the initial variable assignment), and (ii) to drive the search into different regions of the search space (*diversification*), which can also reduce the dependency of the local search on its starting point and thereby make it less dependent on restarts.

Walksat. While the GSAT algorithm and its descendants choose a variable to flip among all the variables that appear in unsatisfied clauses with equal probability, the Walksat strategy by Selman, Kautz and Cohen [134, 93] favors variables that appear in many unsatisfied clauses: It first randomly selects an unsatisfied clause, and then chooses among the variables in that clause. Walksat is a particularly effective local search strategy and its variable selection policy is given in Figure 2.3.

```
proc select-variable(C, A)
      c := a random unsatisfied clause from C
      m := min { b(A, v) : v ∈ c }
      if m = 0  then  s := a variable in v ∈ c with b(A, v) = 0
      else with probability p :     s := a random variable in c
                  probability 1 − p : s := a variable v ∈ c
                                          with minimal b(A, v)
      end
      return s
end
```

Figure 2.3. The Walksat variable selection strategy, ties are broken at random.

Walksat operates as follows. First, it randomly selects one of the clauses from c that are not satisfied by the given assignment A. It then selects one of the variables in c and flips its value, thereby rendering c satisfied. Because of the flip, however, one or more other clauses may become unsatisfied. Therefore, to decide which variable to flip, it considers the number of clauses that become unsatisfied (*break*) when one of the variables in c is flipped, for each of the variables in c.[3] If it is possible to make c true without breaking some other clause, this move is greedily accepted. Otherwise, Walksat follows a probabilistic policy. The entire strategy is given in Figure 2.3, where $b(A, v)$ denotes the number of clauses in C that break if a variable v is flipped given the assignment A.

Incremental Data Structures. The straightforward way of implementing the evaluation of a variable flip (*scoring*) would be to simulate it. However, to efficiently determine which of the many variables leads to the best score, GSAT and its variants employ incremental data structures. An example use is the number of true literals which is recorded and maintained incrementally for every clause. Another is to maintain for every variable the change in the number of satisfied clauses when changing the value of the variable from false to true, assuming the variable is currently false (and vice versa). For a detailed discussion of efficient incremental data structures, see e. g. Michel and Van Hentenryck [107]. Incremental data structures allow the algorithms to move at a rapid pace in the space of variable assignments.

2.3.4 Application Domains of SAT Local Search

Walksat and other flavors of local search for propositional satisfiability have been applied to a variety of combinatorial problems. The following list gives a short survey on application domains various local search procedures for propositional satisfiability have been applied. The domains of study include graph coloring, N-queens and Boolean induction [135, 132], circuit diagnosis, circuit synthesis various planning problems (logistics, rocket, towers of hanoi, blocks world) [134, 93], Sadeh's scheduling problems [35, 148], quasi-groups, and number factorization [54], and Hamiltonian circuit [77]. Some of the above problems have been used in the Beijing SAT competition [101]. Further, SAT local search algorithms have been applied to a collection of benchmarks within the DIMACS effort on "Cliques, coloring, and satisfiability" [90], in particular Boolean function synthesis, circuit testing, parity function learning, randomly generated one-solution problems [128]. Another source of SAT benchmarks which attracted much interest and is *random 3-SAT* [110, 34]. Random 3-SAT has been a driving force for studying and improving local search algorithms e. g. [135, 133, 52, 53, 134, 118].

---

[3] Note that this '*breakcount*' is an approximation of the true change in the number of satisfied clauses.

Maximum Satisfiability. There are also optimization counterparts of SAT, MAXSAT and and weighted and/or partial MAXSAT. The goal in *MAXSAT* is to find an assignment that maximizes the number of satisfied clauses. In *weighted MAXSAT*, a weight is associated with each clause and the sum of the weights of the satisfied clauses is to be maximized. Another recent variant is *partial MAXSAT* in which the clauses are classified into *hard* and *soft* clauses, and the goal is to maximize the number of soft clauses while all hard clauses must simultaneously be satisfied.

Local search methods applied to randomly generated MAXSAT problems have been reported [67, 134, 147], partial MAXSAT has been used for encodings of class-scheduling and random formulas [29], and partial weighted MAXSAT has been used for solving steiner tree problems [85].

2.4 Modeling Languages

The optimization frameworks given above are computational frameworks for *solving* constraint optimization problems that are given by a suitable set of constraints.

Complementary to the aspect of problem solving, *modeling* languages have been designed to allow to formulate constraint problems conveniently and analyze their solutions. An ILP modeling system takes as input a description of the constraints and objective functions in some natural formulation. Often, this representation is quite similar to the mathematical description. Given the model, there is typically some presolving stage performed for simplification and feasibility testing [47]. Subsequently, the model is presented to an ILP solver that is connected to the modeling system. Finally, after the solutions are returned from the solver, the modeling system allows for inspection of the solution. Most modeling systems support a variety of algorithmic codes. The integer local search procedure that will be given later, WSAT(OIP), has been hooked up to the AMPL modeling language [49]. Other popular modeling systems include GAMS [18], LINGO, PLAM [13].

2.5 Search Relaxations and Integer Local Search

Search strategies for combinatorial optimization can be contrasted according to different criteria in order to better understand their characteristics. For example, the 'systematicity' view contrasts search strategies as to whether they are systematic (complete) or follow a local gradient (usually incomplete). For instance, intelligent backtracking and local search can be contrasted in this view and combinations have been proposed [55].

Another view is the 'inference' view, which classifies strategies according to which constraints in a system are treated as primitive (i. e. efficient methods can be used to solve them) vs. non-primitive (i. e. adding them makes the system hard to solve) [144]. Integer programming and finite domain constraint programming can be contrasted and integrated in this view [21].

To improve our understanding of integer local search, let us view combinatorial search in terms of *relaxations*. Combinatorial search means to traverse the nodes of a search space, be it in a systematic or non-systematic way, and independent of the amount of inference that takes place to get from one node to the next. Either way, the essence of combinatorial search is to seek for 'perfect' nodes (in the sense of feasible and optimal solutions) by traversing many 'imperfect' nodes (solutions that meet only a subset of the requirements while violating others).

If the solutions to a discrete optimization problem are defined by variable assignments, the assignments must meet three requirements: (i) *totality*, i. e. all variables have exactly one value assigned, (ii) *integrality*, i. e. all assigned variable values are integral, and (iii) *consistency*, i. e. the assignment meets all problem constraints. We argue here that it is an important factor of a strategy which requirements are *relaxed* during search.

Integer programming branch-and-bound. The relaxation view of integer programming branch-and-bound strategies is: relax the integrality constraints in order to make use of efficient strategies to solve a linear program. In branch-and-bound, this relaxation occurs at each node of a search tree: branching constraints are added, the linear system is solved and all variables are assigned values—some of which happen to be integral. The search tree is traversed in this fashion until a solution is found with all variables integral. Then, the solution is stored, the objective function is bounded and the search continues, in the following using the relaxation bounds for pruning. Hence, the requirement that is relaxed by IP branch-and-bound is *integrality*, while maintaining that all variables be assigned values (*totality*) and all intermediate LP solutions be *feasible* to the problem constraints (one way to define *consistency*).

Constraint satisfaction search. Conversely, complete search strategies for constraint satisfaction (CSP) [142] are based on propagate-and-branch and only assign integral values to variables. Here, the variables range over finite integer domains and the search starts from a *partial* assignment of the variables. At any point, some variables have a value assigned while others are unassigned (more than one value is left in their domain): For instance, suppose we start with a variable $x \in \{1, \ldots, 5\}$, which due to some constraint $x < 3$ becomes $x \in \{1, 2\}$, and next due to branching on x might become assigned to $x = 1$. The power of CSP methods stems from strong propagation algorithms (e. g. arc-consistency) that rule out all variable values that are known to be inconsistent with the current partial variable assignment and the set of constraints (*local consistency*). Search in CSP progresses in the space of partial variable assignments where variable values are speculatively assigned (and possibly later retracted) with the goal to eventually assign one value to *every* variable, such that the solution is optimal. Hence, constraint satisfaction search maintains integrality and local consistency, but relaxes *totality*, i.e. the requirement that all variables be assigned one value.

Integer local search. Complementary to the two previous frameworks, integer local search relaxes the third property: *consistency*. That is, all variables are always assigned individual values, and all values are integral, yet the assignments may violate problem constraints and may thus be *inconsistent*. The goal is to find a consistent solution (one which does not violate any problem constraints) that is optimal. Thus, integer local search maintains integrality and totality, but relaxes consistency.

We hope that this view of combinatorial search can foster the understanding of integer local search and also stimulate ideas for new hybrid methods.

3. Local Search for Integer Constraints

This chapter introduces new local search strategies for integer optimization and represents the technical core of this book. It describes and discusses WSAT(OIP), a domain-independent method that generalizes the Walksat procedure of Selman, Kautz, and Cohen [134] for propositional satisfiability to integer optimization and integer constraint solving. For performance on realistic applications, the method additionally incorporates principles from Tabu Search [58].

WSAT(OIP) operates on *over-constrained integer programs* (OIPs), an algebraic representation for combinatorial optimization problems which is similar to integer programs. The chapter first introduces and discusses OIP, showing that OIP is a special case of integer linear programs. We will argue that OIPs are well suited for devising efficient iterative repair strategies for integer optimization. On the other hand, OIPs allow for combinations with linear programming via a reduction to integer linear programs.

Then the WSAT(OIP) procedure will be described in terms of its basic principles and we give the details of a carefully engineered strategy for the selection of local moves that evolved in the course of the case studies.

Several combinations with linear programming will then be discussed for computing bounds on the optimal solution, approximation and search space reduction. Finally, we will illustrate the basic algorithms and several extensions by a graphical example to review the described techniques. The chapter concludes by discussing related work on local search for integer programming and constraint satisfaction.

3.1 Over-Constrained Integer Programs

The first step in solving an optimization problem is to choose a suitable representation. In Artificial Intelligence, a popular representation is propositional satisfiability (SAT). SAT can represent a variety of interesting combinatorial problems, for example graph coloring [135, 132], circuit diagnosis and synthesis [91, 134], or various AI planning problems [93, 94].

A number of efficient search strategies have been developed for SAT in recent years, both complete [34, 40] and incomplete [135, 134, 64] (the corresponding OR terminology is exact vs. heuristic). However, many combinatorial problems have no concise encoding in propositional logic, especially those involving arithmetic constraints. Hence these algorithms cannot be applied.

For example, consider the pigeonhole problem which occurs as a core problem within many combinatorial problems. Consider a statement with Boolean variables p_{ij}, where p_{ij} means pigeon i is in hole j. A natural encoding is to use two different constraints, (a) every pigeon is in exactly one hole $\sum_j p_{ij} = 1$ (for all i), and (b) no two pigeons are in the same hole $\sum_i p_{ij} \leq 1$ (for all j). Given n pigeons and m holes, this formulation consists of $n + m$ pseudo-Boolean constraints. On the other hand, a SAT encoding would be (a) $\vee_j p_{ij}$ (for all i) and $\forall j \forall k \neq j : p_{ij} \rightarrow \overline{p_{ik}}$ (for all i, $O(m^2 n)$ clauses); similarly for (b). With $O(m^2 n + n^2 m)$ constraints, the size of the SAT encoding would be impractical for larger instances.

Pseudo-Boolean / 0-1 Integer Constraints. The class of *linear pseudo-Boolean constraints* (linear 0-1 integer constraints) is defined as follows [66]. A linear pseudo-Boolean constraint is of the form

$$\sum_{i \in I} c_i \cdot L_i \sim d, \tag{3.1}$$

where c_i, d are rational numbers, $\sim \in \{=, \leq, <, \geq, >\}$, and the L_i are literals for all $i \in I$ (a literal is a Boolean variable or its negation). A set of pseudo-Boolean constraints together with an objective function yields a 0-1 ILP. Pseudo-Boolean constraints generalize SAT in the sense that every Boolean clause (disjunction of literals) can be translated into a single linear pseudo-Boolean inequality. For example, the clause $\overline{x} \vee y$ would be translated to $(1 - x) + y \geq 1$.

On the other hand, when converting linear pseudo-Boolean inequalities to propositional satisfiability, the number of SAT clauses required to represent one inequality can grow exponentially with the number of variables in the inequality. To see this, consider an example from Barth [12], the conversion of $l_1 + \cdots + l_n \geq d$. Without introducing new variables, its equivalent SAT representation is a conjunction of $\binom{n}{n-d+1}$ SAT clauses,

$$\bigwedge_{I \subseteq \{1,\ldots,n\}:|I|=n-d+1} (\vee_{i \in I} l_i).$$

Generalizing SAT Local Search. There are still practical shortcomings of the pure pseudo-Boolean constraint representation (3.1). First, optimization problems (as opposed to decision problems) contain objectives which should be included in the representation.

Second, certain problems may better be represented using non-Boolean decision variables (e. g. production of a good could vary between 1 and 100 items). Compiling such problems to a pseudo-Boolean system would again incur an increase in both the number of variables and the length of the constraints.[1] Depending on the applied algorithms, the structure present in the original formulation may not be accessible in the compiled version of the problem.

Despite the fact that SAT does not quite have the right expressivity for many realistic optimization problems, efficient methods exist for SAT and it is desirable to generalize their principles to more expressive constraint classes. Obtaining leverage from a generalization is especially interesting for the efficient SAT local search algorithms developed recently, e.g. Walksat [134].

Over-constrained Integer Programs: Motivation. To give a natural generalization, we will introduce *over-constrained integer programs (OIPs)*, an algebraic representation that is suited for a range of combinatorial optimization problems. With respect to expressivity, we will show that OIPs are a special case of integer linear programs (ILPs).

The principal difference to ILPs is that while ILPs use a monolithic objective function, OIPs represent the overall optimization objective by many competing soft-constraints. Using soft constraints to encode objectives, it is natural to apply *iterative repair* [109] to integer optimization. Further, the OIP structure can be exploited by iterative local search: After a short initial phase, only a small fraction of the soft constraints are violated and the search can focus on repairing those violated constraints.

In general, an over-constrained system is a set of constraints in which typically not all constraints may be satisfied and some are therefore marked as 'soft' or ranked according to a preference system [84]. Optimization objectives have also been encoded by soft constraints in the context of constraint hierarchies [22].

OIPs are similar to ILPs. So can we make use of techniques from linear programming? The answer is yes. We will give a transformation of OIP into ILP that is inspired from piecewise-linear programming. Another aspect concerns standard modeling languages like AMPL [48]. Can they be applied to model OIPs? The answer is, again, yes. ILP modeling languages can be used for modeling. In fact, combining a modeling language with a local search algorithm directly yields a practical constraint solver. For illustration, we will provide a detailed AMPL model of one case study from sports scheduling in the Appendix.

3.1.1 Definition

We refer to a constraint system of hard and soft inequalities over finite domain variables as an *over-constrained integer program, (OIP)*. Here, we only consider the case where all constraints are linear inequalities and the system is given by a tuple $\mathcal{O} = (A, \mathbf{b}, C, \mathbf{d}, \mathbf{D})$, formulated in matrix notation as

---

[1] Binary and gray code representation for large variable domains are possible [105], but we will see some limitations for local search in Chapter 7.

$$Ax \geq b$$
$$Cx \leq d \quad (soft) \qquad (3.2)$$
$$x \in D,$$

where $A = (a_{ij})$ and $C = (c_{ij})$ are $m \times n$ coefficient matrices, b and d are m-vectors, and $x = (x_1, \ldots, x_n)$ is the variable vector, ranging over positive finite domains $D = (D_1, \ldots, D_n)$.[2] (3.2) will be interpreted as the OIP minimization problem

$$\min\{ \|Cx - d\| \ : \ Ax \geq b, \ x \in D\}, \text{ with } \|v\| := \sum_i \max(0, v_i). \quad \text{(OIP)}$$

In $\|.\|$, the contribution of each violated soft constraint to the overall objective is its degree of violation. An assignment to all the variables that satisfies all hard constraints of $\mathcal{O}$ is called a *feasible solution*, and for every feasible solution s, the value of $\|Cs - d\|$ will be called *soft constraint violation* of $\mathcal{O}$. As will be shown, OIPs are a special case of integer linear programs.

Based on OIPs, a single strategy will later be formulated both to find feasible solutions to difficult problems of hard constraints, and to find good solutions to optimization problems. The strategy will proceed by iteratively repairing violated hard and soft constraints.

3.1.2 Relation to Integer Linear Programs

For many practical applications, we would like to enable collaboration of OIP solvers with existing optimization technology based on linear programming. Examples of such combinations are lower bounding techniques (like linear relaxations or Lagrangean relaxation), approximation algorithms based on linear programming, or search space reduction techniques (see Section 3.3 for a detailed discussion).

Due to the evaluation $\|.\|$, however, a given OIP is not immediately equivalent to an integer *linear* program (ILP). Next, it will be shown that every OIP can be converted to an ILP. We consider the following cases: (i) If a given OIP has a certain property (called *confinedness*), it is equivalent to an ILP. Further, (ii) every OIP can be converted to an equivalent ILP.

Equivalence is defined here in the sense that two ILPs/OIPs are equivalent if they have the same feasible solutions and their objective function values are the same for every feasible solution.

---

[2] For conciseness, we will discuss OIPs in the form (3.2) which could be referred to as *min-normal form*. Every OIP minimization problem can be converted into min normal form by multiplying every incorrectly directed inequality (e. g. $\leq$ instead of $\geq$) by -1 and converting every equality into two inequalities.Input to the algorithms described in the following is not required to be in min-normal form. Also, implementations use sparse matrix representations.

Notational Convention. To simplify the following discussion, we will write

$$\min\{cx \; : \; Ax \geq b, \; x \in D\},$$

as an abbreviation for the ILP

$$\min\{cx' \; : \; Ax' \geq b, \quad x_i' = \sum_{j \in D_i} j \cdot y_{ij}, \; \sum_{j \in D_i} y_{ij} = 1, \; \; y_{ij} \in \{0,1\}\}, \quad (3.3)$$

in which the finite domain variables $x_i \in D_i$ are replaced by variables x_i' and additional binary variables $y_{ij} \in \{0,1\}$ are introduced for all $j \in D_i$. If all finite domains range over integers (no loss of generality) then (3.3) is an ILP $(x \in \mathbb{Z}_+^n)$.

Definition 3.1.1 (Confinedness). *An over-constrained integer program* (A, b, C, d, D) *is confined if and only if for every feasible solution* s, *the following holds:*

$$Cs - d \geq \begin{pmatrix} 0 \\ \vdots \\ 0 \end{pmatrix}.$$

Proposition 3.1.1. *A confined OIP* (A, b, C, d, D) *is equivalent to the ILP*

$$\min\{ \; \|Cx - d\|_1 \; : \; Ax \geq b, \; x \in D\}, \; with \; \|v\|_1 := \sum_i v_i, \quad (3.4)$$

Proof. As confinedness holds, $\|.\|$ is equivalent to the norm $\|.\|_1$ for all feasible solutions. □

The following proposition shows that every OIP can be converted to an ILP by introducing additional constraints.

Proposition 3.1.2. *An over-constrained integer program* $(A, b, (c_i), d, D)$,

$$Ax \geq b, \quad c_i x \leq d_i \; (soft), \quad x \in D$$

is equivalent to the ILP

$$\min\{\sum_i e_i \; : \; Ax \geq b, \; c_i x + s_i - e_i = d_i, \; e_i, s_i \geq 0, \; x \in D\}. \quad (3.5)$$

Proof. The basic idea behind the conversion is that every soft inequality can be converted into an equality constraint with two additional variables: a slack variable s_i and an excess variable e_i, where the excess variable accounts for the incurred penalty. The OIP minimization problem is

$$\min\{\sum_i \max(0, c_i x - d_i) \; : \; Ax \geq b, \; x \in D\}.$$

To show equivalence, it suffices to show equality of the two objective functions for all $\mathbf{x}$. We will employ (componentwise) ordered tuples $\langle a, b \rangle$. It remains to show that for all $\mathbf{x}$,

$$\min\{\langle \mathbf{x}, e_1 + \cdots + e_m \rangle \ : \ \mathbf{c}_i\,\mathbf{x} + s_i - e_i = d_i, \ \ s_i \geq 0, e_i \geq 0 \}$$
$$= \langle \mathbf{x}, \max(0, \mathbf{c}_1\,\mathbf{x} - d_1) + \cdots + \max(0, \mathbf{c}_m\,\mathbf{x} - d_m) \rangle.$$

This can be shown componentwise for all i because the s_i, e_i are independent from s_j, e_j for all $i \neq j$. For each i, we consider the components $\langle \mathbf{x}, e_i^* \rangle = \min\{\langle \mathbf{x}, e_i \rangle \ : \ldots\}$ and $\langle \mathbf{x}, \max(0, \mathbf{c}_i\mathbf{x} - d_i) \rangle$. We consider two cases:

1. $\mathbf{c}_i\,\mathbf{x} - d_i < 0$. The right-hand-side is $\max(0, \mathbf{c}_i\,\mathbf{x} - d_i) = 0$. The left-hand-side also yields $e_i^* = 0$ since this is the smallest $e_i \geq 0$ satisfying $e_i = \mathbf{c}_i\,\mathbf{x} + s_i - d_i$.
2. $\mathbf{c}_i\,\mathbf{x} - d_i \geq 0$. Now the left-hand-side yields

$$e_i^* = \min\{e_i \ : \ e_i = \mathbf{c}_i\,\mathbf{x} + s_i - d_i, s_i \geq 0, e_i \geq 0\} = \mathbf{c}_i\,\mathbf{x} - d_i,$$

which is also the value of the right-hand-side. $\qquad\qquad\qquad\qquad\square$

In summary, OIPs are a special case of ILPs. We notice that all OIP models which will be given in the case studies happen to be confined, which is typically easy to check.[3] The algorithms that will be described subsequently are directly applicable to general OIPs, however.

There is a close relation of OIP to integer linear programs with piecewise-linear objective functions.[4] A piecewise-linear function is a function that is pieced together from linear segments. Every soft constraint $\mathbf{cx} \leq d$ gives rise to a penalty that is expressed by a piecewise-linear convex function $\max(0, \mathbf{cx} - d)$, as shown in Figure 3.1. Given the fact that the sum of two convex functions is a convex function, OIP objectives are piecewise-linear convex. Piecewise-linearities are employed in models to give a more realistic description of costs than can be achieved by linear terms alone [48]. There are also extensions of the simplex algorithm for piecewise-linear convex programming [46].

An OIP Example. As an example of a piecewise-linear objective function, consider the problem of assigning a set of tasks T to a workforce W. Suppose that every task consumes one hour and every worker w is employed for 8 hours, but can work overtime at some cost. The problem can be modeled with binary variables, $a_{tw} = 1$ iff task t is assigned to worker w. The constraints are:

---

[3] Note that the following is a simple sufficient condition for confinedness: for every soft constraint $\mathbf{cx} \leq d$, there exists a hard constraint $\mathbf{cx} \geq d$.

[4] I thank Andrew Parkes for bringing my attention to the relation between soft constraints and piecewise-linear functions, which eventually lead to Proposition 3.1.2.

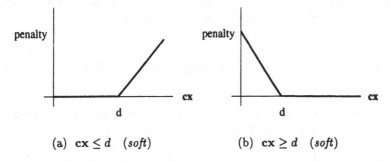

Figure 3.1. Piecewise-linear penalty functions

Assign every task	$\sum_w a_{tw} = 1$, for all t.
Limit workday	$\sum_t a_{tw} \leq 12$, for all w.
Minimize overtime	$\sum_t a_{tw} \leq 8$ (*soft*), for all w.

The goal is to assign the tasks to the workers such that the summed overtime is minimized. The problem as stated above is not a hard combinatorial problem but it may occur as subpart of one.

3.1.3 Constraint-Bounds

A soft constraint $\mathbf{cx} \leq d$ of an over-constrained integer program is similar to an objective function with a bound. Therefore, the right-hand side d will be referred to as *constraint-bound*. For modeling, we need to understand the degrees of freedom of choosing the constraint bounds: Consider an OIP $\mathcal{O}$ containing a soft constraint $s : \mathbf{cx} \leq l$, to minimize the excess of a function $\mathbf{cx}$ over l. We observe that the larger the values of l, the larger the (soft) feasible region of s. Sometimes, for a given value of l there may be no feasible solution to the problem that satisfies s. This raises the question in what range the constraint-bounds can be increased or decreased without changing the optimization problem?

Definition 3.1.2 (Rebounding). *Given an OIP $\mathcal{O} = (A, \mathbf{b}, C, \mathbf{d}, \mathbf{D})$, the OIP $(A, \mathbf{b}, C, \mathbf{d'}, \mathbf{D})$ is a called a rebounding of $\mathcal{O}$.*

Proposition 3.1.3 (Invariance under Confined Rebounding). *If $\mathcal{O}_1$ and $\mathcal{O}_2$ are confined OIPs, and $\mathcal{O}_2$ is a rebounding of $\mathcal{O}_1$, then the set of optimal solutions is the same for $\mathcal{O}_1$ and $\mathcal{O}_2$.*

Proof. Since $\mathcal{O}_1$ is confined, it is equivalent to an IP with minimization objective $\|C\mathbf{x} - \mathbf{d}_1\|_1$. As $\mathcal{O}_2$ is also confined and a syntactic rebounding of $\mathcal{O}_1$, its corresponding IP has the same set of constraints but a minimization objective of $\|C\mathbf{x} - \mathbf{d}_2\|_1 = \|C\mathbf{x} - \mathbf{d}_1\|_1 + k$, for some k. Obviously, every solution $\mathbf{x}$ of $\mathcal{O}_1$ with objective value s is a solution of $\mathcal{O}_2$ with objective $s + k$ and thus the sets of optimal solutions are the same.   □

Proposition 3.1.3 implies that if for operational purposes it is helpful to increase the constraint-bounds, this may be done provided that confinedness is maintained.

Proposition 3.1.4. *Let* $(A, \mathbf{b}, C, \mathbf{d}, \mathbf{D})$ *be a confined OIP. For every* $\mathbf{d}'$ *with* $d_i' \le d_i$ *for all* i, *the rebounding* $(A, \mathbf{b}, C, \mathbf{d}', \mathbf{D})$ *is confined.*

The proof follows directly from the definition of confinedness. Together with Proposition 3.1.3, Proposition 3.1.4 states that optimal solutions do not change if the soft constraint-bounds are tightened (note that their objective function values do, however). This tells us that choosing soft constraint-bounds too tightly does not affect the optimization problem except for a shift in the objective function values.

3.2 Integer Local Search: WSAT(OIP)

This section introduces WSAT(OIP), a local search method to solve optimization and feasibility problems represented by OIPs. Starting from some initial assignment, the procedure performs changes of variable values, thereby moving in the space of integer assignments to find feasible or good solutions.

WSAT(OIP) generalizes Walksat (described in section 2.3), but its performance on realistic problems is often critically dependent on the incorporated concepts from Tabu Search [61]. As the case studies will demonstrate, despite its conceptual simplicity, WSAT(OIP) is surprisingly effective in terms of performance and robustness. Historically, WSAT(OIP) builds upon WSAT($\mathcal{PB}$) [149] which performs stochastic tabu search on over-constrained pseudo-Boolean systems. Generalizing from Boolean variables to general finite domain variables, WSAT(OIP) subsumes WSAT($\mathcal{PB}$).

The Strategy. We describe the method starting from its basic principles, and proceed by giving the main loop of the algorithm and the (important) details of how the local moves are selected. A summary of the parameters concludes the description. Section 3.4 will further illustrate the search process and several extensions by a graphical example.

WSAT(OIP) follows an 'iterative repair' strategy and operates on a *total assignment* (an assignment to all variables). Individual variable/value pairs are iteratively selected to be changed in order to improve the local gradient of an overall measure of the satisfaction of the constraints. The main cycle for selecting local moves is illustrated in Figure 3.2. Generalizing from the Walksat algorithm [134], variable changes in WSAT(OIP) are selected in a two-stage strategy of first randomly selecting an *unsatisfied* constraint for partial

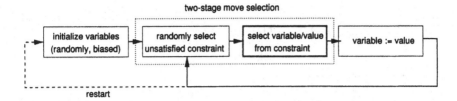

Figure 3.2. Local search and the two-stage control strategy of Walksat.

repair and from the constraint selecting a variable to be changed.[5] This two-stage control strategy, which we will call '*Walksat-Principle*', distinguishes Walksat among the many flavors of recent local search algorithms for SAT and CSP (including GSAT [135], GSAT +walk [134] and MIN-CONFLICTS [109]).[6] Note that it favors those variables that appear in many unsatisfied constraints [134].

The criterion for move selection is to perform hill-climbing on a *score* which reflects both the degree of infeasibility and the optimization objective. A value change of a Boolean variable is to *flip* (complement) the variable to improve the score. As a generalization, a move of WSAT(OIP) consists of *triggering* the value of a finite domain variable to a smaller or greater value in the neighborhood of its current value assignment. Occasionally, a restart with a new initial assignment takes place to escape from local optima, for example after a fixed number of moves.

3.2.1 The Score

To describe the move selection strategy for over-constrained IPs in detail, we first need a score definition. Given a particular assignment $\mathbf{x}$, a system of the form (3.2) is evaluated as,

$$score(\mathbf{x}) = \|\mathbf{b} - A\mathbf{x}\|_\lambda + \|C\mathbf{x} - \mathbf{d}\|. \tag{3.6}$$

using the usual norm $\|\mathbf{v}\| = \sum_i \max(0, v_i)$. A useful property of the score (3.6) is that it is identical to the value of the objective function of the equivalent ILP.

Additionally, the score (3.6) uses a vector $\lambda \geq 0$ for weighting the hard constraint violations, defined by $\|\mathbf{v}\|_\lambda := \sum_i \lambda_i \max(0, v_i)$. Note that the soft constraints do not carry weights in order that the score of feasible solutions be *equal* to the objective function value. Only one case study will make use of non-unit weights, and all experiments were performed with weights that were assigned statically.

---

[5] In contrast to SAT, a variable change does not always repair the selected constraint completely.

[6] More generally, this two-stage control strategy is (i) select a constraint c for repair, and (ii) select a partial repair for constraint c.

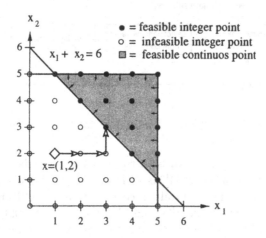

Figure 3.3. Manhattan distance, a shortest path to enter the feasible region of a constraint with all coefficients from $\{-1, 0, 1\}$.

Graphical Illustration. To illustrate the score (3.6), assume a current variable assignment **s** and a constraint $c : \mathbf{cx} \geq d$ with all elements of **c** from $\{-1, 0, 1\}$. Then, $\|d - \mathbf{cs}\|$ measures the number of unit variable changes required to enter the feasible region of the constraint, i.e. the *Manhattan distance* of **s** to c. Figure 3.3 illustrates the situation for the constraint $x_1 + x_2 \geq 6$ over two variables $s_1, s_2 \in \{1, 2, \dots, 5\}$. The score of $\mathbf{s} = (1, 2)$ is $6 - (1 + 2) = 3$ unit variable changes. Section 8.2 will discuss a refinement of this scoring scheme.

3.2.2 The Main Loop

In each iteration, WSAT(OIP) makes a change of exactly one variable-value pair, Figure 3.4 illustrates the basic main loop of WSAT(OIP). If an assignment is found that is better than the best one found in the past, this new best assignment is stored. If an assignment is found that is known to be optimal (e. g. using a relaxation-proof), the optimal assignment is returned. A restart is performed after a fixed number of iterations.

3.2.3 Move Selection and Tabu Search Extensions

The fundamental principle behind WSAT(OIP) is steepest-descent (selecting local moves that most improve the total score) combined with adaptive memory [58] and a noise strategy to overcome local minima. The remaining degrees of freedom are how to select (i) a constraint and (ii) a partial repair, i.e. a variable and its new value. It a time-consuming engineering task to find good strategies for (i) and (ii) which has a strong impact on performance. The

proc WSAT(OIP)
    input OIP $\mathcal{O}$, Maxmoves, Maxtries
    output an approximately optimal feasible solution
        for $\mathcal{O}$, if found
    for $i := 1$ **to** Maxtries **do**
        $a :=$ initial total assignment, $a[v] \in dom(v)$,
            possibly infeasible
        $u := \infty$
        for $j := 1$ **to** Maxmoves **do**
            if a is known to be optimal **then return** a
            if a is feasible $\wedge$ $score(a) < u$ **then** $u := a$
            $c := select\text{--}unsatisfied\text{-}constraint(\mathcal{O}, a)$
            $\langle v, s' \rangle := select\text{--}partial\text{-}repair(\mathcal{O}, c, a)$
            $a := a[v \leftarrow s']$
        end
    end
    if $u < \infty$ **then return** u
            else return "no feasible solution found"
end

Figure 3.4. Main loop of WSAT(OIP) for over-constrained integer programs.

influence of different variable selection strategies on performance has been investigated previously for SAT local search [118, 106]. Throughout the case studies, unsatisfied constraints are selected at random. Although different constraint selection schemes have been studied (among those selecting one of the violated constraints ordered by increasing/decreasing 'constrainedness'), we could not find a selection scheme that improved over random selection.

Figure 3.5 gives a scheme that combines successful elements from stochastic local search and deterministic tabu search. The particular strategy evolved in the course of our case studies and includes a tabu mechanism, history-based tie-breaking, and timid noise: A tabu mechanism with tenure of size t avoids assigning a variable-value pair that has been assigned in the previous t moves—unless the score would improve over the best past score (in tabu search terminology, overriding the tabu-status of a move is called *aspiration*). All ties between otherwise equivalent variable-value pairs are broken by a history mechanism inspired by Gent and Walsh's HSAT [53]: On ties, choose the move that was chosen i) least frequently, and then ii) longest ago.

In contrast to SAT local search, the length of 0-1 inequalities that can be handled efficiently can be quite large. This is because only those variables need to be scored that contribute to the violation of a constraint. For example, in the course assignment problems, inequalities contained up to 350 variables of which typically only 10% needed to be scored for repair.

1. Randomly select an unsatisfied constraint α (with probability p_{hard} a hard constraint, and with $1 - p_{hard}$ a soft constraint).
2. From α, select all variables which can be changed such that α's score improves. For each such variable, select one or more α-improving values and compute the hypothetical total scores (Boolean variables are flipped, finite domain variables are triggered up or down by at most d stpes).
3. From the selected variable-value pairs, remove the ones which are *tabu* (tabu-aspiration by score).
4. Of the remaining variable-value pairs, select one that most improves the total score, if assigned. Break ties according to i) *frequency* and ii) *recency*.
5. If the total score cannot be improved: With probability p_{noise}, select a random α-improving non-tabu variable-value pair. With $1 - p_{noise}$, select the best possible one.

Figure 3.5. A stochastic tabu search strategy for move selection in WSAT(OIP). The strategy extends Walksat to systems of hard and soft constraints (1.) and its variable selection to general finite domain variables (2.). Additionally, the straegy is merged with adaptive memory from tabu search (3.) and an extended version of history-based tie-breaking (4.) [53].

Parameters. Finally, Table 3.1 summarizes the parameters of the basic algorithm. It also reports parameter ranges as rules of thumb. Most parameter settings in the case studies were performed with settings from these ranges (the exact values will be reported). Additionally, for practical purposes a seed parameter is used: fixing the seed yields a deterministic algorithm.

Table 3.1. Parameters of WSAT(OIP) with standard ranges.

parameter	std range	description
p_{noise}	0–0.2	probability of a allowing random downhill move
p_{hard}	0.8–0.9	probability of selecting a hard constraints for repair
p_{zero}	0.5–0.9	probability of initializing a variable with zero
t	1–2	tabu-tenure (overridden by score-aspiration)
Maxstep	2	maximal trigger distance of non-binary variables
tie-breaking	on	history-based tie breaking (frequency,recency)
Maxtries		total number of tries
Maxmoves		number of moves within a try

3.3 Combinations with Linear Programming

Local search methods like WSAT(OIP) suffer from several principal drawbacks when used in an optimization context. These are:

(i) The inability to detect if a (specific) obtained solution is optimal or to estimate how far from optimality the solution is.
(ii) The lack of any guarantee of the quality of the solutions returned. (For some NP-hard problems, approximation algorithms exist that can give such, although weak, guarantees [73]).
(iii) The absence of sound search-space pruning methods during the local search process.

We will argue that combinations with linear programming can help to overcome these drawbacks to some extent. Especially, for (i) lower bounding, (ii) initial assignment by rounding, and (iii) search space reduction.

The first section describes the standard use of the linear relaxation to obtain (lower) bounds for (minimization) problems. The second section describes the computation of initial assignments by linear programming followed by randomized rounding, and a novel extension of this idea for WSAT(OIP). And the third section discusses the use of 'reduced costs' for search space reduction, as proposed by Balas and Martin [9].

3.3.1 Bounds from LP Relaxations

It is well-known that linear programming problems are polynomially solvable in theory, and can often be solved efficiently in practice [31]. This fact can be exploited by using linear programming (LP) to efficiently compute bounds on the optimum of a problem at hand. In practice, bounds obtained directly from the LP relaxation are often valuable estimates of the solution quality.

Consider the following 0-1 integer problem in matrix notation (note that LP relaxations are applicable to general integer programs)

$$
\begin{aligned}
Z_{\mathsf{IP}} = \text{minimize } & \mathbf{cx} \\
\text{subject to } & A\mathbf{x} \geq \mathbf{b} \\
& x_i \in \{0, 1\}.
\end{aligned}
\tag{3.7}
$$

One way to generate a lower bound to problem (3.7) is to relax the integrality constraints $x_i \in \{0, 1\}$ by substitution with bounds on the variables. This yields the linear program

$$
\begin{aligned}
Z_{\mathsf{LP}} = \text{minimize } & \mathbf{cx} \\
\text{subject to } & A\mathbf{x} \geq \mathbf{b} \\
& x_i \in [0, 1].
\end{aligned}
\tag{3.8}
$$

As (3.8) is a relaxation of (3.7), every lower bound of (3.8), in particular its optimal solution, is a valid lower bound of (3.7). In absence of the optimal objective value for the integer programming problem, this fact can be used to estimate the quality of a solution to (3.7) by solving the linear relaxation (3.8).

The difference between the optimal IP and LP solutions, $Z_{IP} - Z_{LP}$, is usually referred to as *integrality gap*. Practitioners have observed that problems with small integrality gaps tend to be solved more efficiently by integer programming branch-and-bound than problems with large gaps. Other general methods for lower bounding could be used as well, e.g. Lagrangean relaxation [17].

3.3.2 Initialization by Rounding LP Solutions

Solving the linear program followed by rounding the non-integral solutions can be combined with integer local search in useful ways. Consider a combinatorial optimization problem given by the ILP

$$\begin{aligned} \text{minimize} \quad & \mathbf{cx} \\ \text{subject to} \quad & A\mathbf{x} \geq \mathbf{b} \\ & x_i \text{ integer} \end{aligned} \qquad (3.9)$$

As before, the first step is to relax the integrality constraints and apply linear programming to solve the system

$$\begin{aligned} \text{minimize} \quad & \mathbf{c\hat{x}} \\ \text{subject to} \quad & A\hat{\mathbf{x}} \geq \mathbf{b}. \end{aligned} \qquad (3.10)$$

The result is an LP-optimal solution $\hat{x}_i, i = 1 \ldots n$. As the $\hat{x}_i$ may be fractional, the LP solution may not constitute a feasible solution for the integer program. In order to restore integrality, the resulting values $\hat{x}_i$ can therefore be rounded (i) deterministically or (ii) in a randomized fashion (up or down) to yield values $\bar{x}_i$. There are several benefits that can be obtained from rounding, and we will start by describing two practical benefits that arise irrespectively of the employed rounding scheme. Subsequently, we will touch on a theoretical benefit of the combination.

Practical Benefits. Some discrete optimization problems exhibit a low level of discreteness because even although all variables are integer, their domains are large. An efficient way to solve such problems is by linear programming and rounding in the initialization stage of an integer local search method. To illustrate this, consider the following problem formulation, occurring as a subtask of a real configuration problem.[7]

---

[7] Thanks to Thomas Axling for providing the example.

$$Z_{\text{IP}} = \text{minimize} \quad x_1 + 20x_2 + 20x_3 + 20x_4$$
$$\text{subject to} \quad x_1 + 4x_2 + 2x_3 + x_4 \geq 800$$
$$x_1 + x_2 + 4x_3 + x_4 \geq 100$$
$$x_1 + x_2 + x_3 + 4x_4 \geq 600 \quad\quad (3.11)$$
$$x_1 + x_2 + x_3 + x_4 \geq 400$$
$$x_i \in \{1, \ldots, 1000\}.$$

Due to the large variable domains $\{1, \ldots, 1000\}$, this problem is in a sense 'close' to its linear relaxation. This fact is illustrated by Table 3.2 which contains the optimal LP and IP solutions. Using the optimal LP solution and

Table 3.2. Optimal solutions for a subtask of a configuration problem (3.11).

var	LP	IP
x_1	200	201
x_2	133.33	133
x_3	0	0
x_4	66.667	67
z	6000	6010

deterministically rounding it yields a solution which violates only the first constraint in problem (3.11).

Because of the type of local moves, WSAT(OIP) as described above is not the method of choice for problems with very large domains if started from a random initial variable assignment. In fact, in this particular example, it behaves poorly. However, when initialized with the rounded LP solution from Table 3.2 it only takes one variable flip to restore feasibility and arrive at the optimal IP solution. This is because examining the ways to repair the only violated constraint, all variables of the first constraint are scored and increasing x_1 yields the smallest increase in the overall score.

Observations for Binary Variables. Even if all variable domains are binary, rounding the non-integral solutions obtained from linear programming can be helpful. In general, feasibility with the constraints is lost due to the rounding step. However, integer local search can be expected to recover feasibility quickly in most cases. An additional benefit may arise with problems for which the LP relaxation yields a large number of integral values. For instance on "large-sized" generalized assignment benchmarks from OR-library [16], instances exist for which the LP optimum has only 1% (5 out of 500!) non-integral variables; all other variables were 0 or 1.

Leashed Local Search. For the case that a significant number of integral values are obtained from the LP relaxation, we propose an extension of WSAT(OIP) which we call *leashed local search*. Starting from an initial rounded LP solution, perform local search as usual. However, limit the radius the search may divert from the initial assignment, by limiting the hamming distance between the initial solution and the current solution (the *hamming distance* between two Boolean variable assignments is the number of bits that differ). We did not explore this technique in the case studies because the principal aim of this work is to examine what can be achieved by local search; the effectiveness of Leashed Local Search on the other hand would be tied to the question how close to the IP solution a corresponding LP solution is. Also, in some of our case-studies (see Section 6.1), solving the LP relaxation took orders of magnitude longer than solving the original integer problem with WSAT(OIP).

Approximation by Randomized Rounding. The approach of randomized rounding is due to Raghavan and Thompson [122], and can be used to formulate efficient (polynomial) approximation algorithms with provable performance guarantees. The key insight of randomized rounding is that certain performance guarantees can be derived if one solves a linear relaxation (3.11) and randomly rounds the fractional variables. Generally, if $\hat{x}_i$ is the solution obtained from linear programming, we assign a solution variable for the IP, $\bar{x}_i = 1$ with probability $\hat{x}_i$ and $\bar{x}_i = 0$ with probability $1 - \hat{x}_i$. Recently, a number of approximation algorithms for *NP*-hard problems have been presented based on randomized rounding, whose characteristic is to give a quality guarantee for the returned solution [73]. One example is maximum satisfiability (MAXSAT) for which randomized rounding can be combined with random variable assignment ($p_{zero} = 0.5$) to yield an approximation algorithm which guarantees that its solution satisfies at least 3/4 of the clauses [154, 112].

If, for a particular problem, such a performance guarantee exists for randomized rounding, this guarantee is of course directly inherited by an integer local search method if randomized rounding is used to obtain the initial solution.

3.3.3 Search Space Reduction Using LP Reduced Costs

We next describe a method for dynamic search space reduction of local search which can be employed in combination with WSAT(OIP), and which has been reported by Balas and Martin [9] and Abramson *et al.* [4].

The idea is that solving the LP relaxation of an integer program to optimality reveals information about the sensitivity of the solution with respect to changes in the problem's parameters. In mathematical programming, such analysis is referred to as *sensitivity analysis* [31, 152].

An important instrument in sensitivity analysis are *reduced costs*: When solving the linear program to optimality using the simplex method, together with the optimal solution one obtains an optimal tableau which contains basic variables (the variables that are non-zero in the optimal solution) and nonbasic variables (the variables that are zero). We describe the idea in terms of 0-1 integer programs.

For any nonbasic variable v_i, its reduce cost r_i is the amount by which v_i's objective function coefficient must be improved before v_i will become a basic variable in some optimal solution to the LP. Another interpretation of r_i is that it is the amount by which the objective function Z_{LP} will increase (in a minimization problem) if the variable is increased by one.

Reduced costs can be employed for a dynamic pruning of the search space in the following way. If at any time during the search the best feasible solution found (upper bound) has an objective function value of Z_{UB}, then all variables v_i can be fixed to zero for which $Z_{LP} + r_i > Z_{UB}$. This is true because assigning one to these variables would yield a solution which is provably worse than the best solution already found.

This strategy can be used to extend WSAT(OIP) by allowing it to fix variables in the following way: Every time an improved feasible solution (upper bound) with objective value Z_{UB} is found, all variables v_i for which $Z_{LP} + r_i > Z_{UB}$ holds can be fixed to 0. Notice that the current value of v_i must then be 0 (because if $v_i = 1$ then it would follow that $Z_{UB} \geq Z_{LP} + r_i$ which is in contradiction with $Z_{UB} < Z_{LP} + r_i$). If all variables are fixed, the upper bound is provably optimal.

A requirement for the pruning to be effective is that a significant percentage of the reduced cost values be sufficiently large. For example, this is typically the case for set-partitioning problems where many variables can be fixed by the technique.[8] In the case studies conducted here, except for Chapter 7, the reduced costs were not significant (the particular covering and assignment problems had mostly zero reduced costs and the feasibility problems have no true reduced costs because there is no objective function).

Preliminary Results. Because of the small reduced costs in our case studies, we have not evaluated search space reduction in detail at this point. Nevertheless, in addition to the observations by Abramson *et al.* [4], our observations on generalized assignment benchmarks from OR-library [16] are very promising. For example, on a 'large-sized' GAP instance, 334 out of 500 binary variables would have been fixed to 0 during the search process using the upper bound found by WSAT(OIP) after a short time—a remarkable reduction of the search space.

3.3.4 Implementation Issues

WSAT(OIP) has been implemented in C/C++, making use of the Lex/Yacc compiler generator for parsing and of the standard Gnu C++ container libraries for hashing of variable names. Including interfaces to Oz, AMPL, and

---

[8] David Abramson, personal communication.

CPLEX the code is roughly 7000 lines of source code. However, the core algorithms of WSAT(OIP) require only around 2500 lines, and thus represent a comprehensible piece of code.

Incremental Data Structures. In order to perform the local moves efficiently, it has often been emphasized for SAT local search that incremental data structures can greatly enhance the performance. The fundamental data structure used to represent the unsatisfied constraints are two linked arrays (used once for hard and once for soft constraints): At any point in time, one direction $u[1 \ldots k]$ maintains the indices of all k unsatisfied constraints for efficient selection of a random unsatisfied constraint. The other direction $ui[1 \ldots m]$ is used to efficiently update u and maintains the index where constraint i is positioned in the u array if currently unsatisfied, i.e. $u[ui[i]] = i$, illustrated in Figure 3.6. These arrays are updated whenever a constraint changes between sat and unsat.

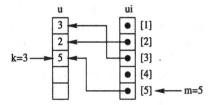

Figure 3.6. Linked arrays for efficient constraint selection and updates. Here, three constraints (3,2,5) are unsatisfied.

The bottleneck computation of the local search process, however, is the evaluation of the proposed moves. Therefore, the WSAT(OIP) implementation employs an extension of the incremental data structures used by Walksat and GSAT [135], which have been described in detail in [107].

The evaluation of the left hand side of each constraint under the current assignment ($\mathbf{cx}$) is incrementally maintained. Further, to efficiently compute a change of the overall score upon changing a variable value, a list is maintained for each variable v which links v to all the constraints it occurs in with non-zero coefficient. To compute the score of changing a variable, it is sufficient to visit only the constraints that it appears in, add or subtract the local coefficient to the current left hand side, and reevaluate if the constraint becomes sat or unsat. After changing the value of a variable, an update takes place of the u and ui arrays, and the value of the incremental total score is adjusted as before.

Another important feature of the algorithmic design for constraints with many variables is that only those variables are 'scored' for which changing can improve the score of the *selected* constraint. This turned out to be a critical extension in a number of case studies with large numbers of variables per constraint.

3.4 A Graphical Interpretation

To illustrate the search process, its limitations and various ways to improve it, we will now look at how WSAT(OIP) moves in the space of integer solutions in a graphical example. Consider the following OIP (3.12) and its graphical equivalent in Figure 3.7, as it is usually depicted for IP/LP problems.

$$
\begin{array}{lll}
\text{(A)} & 9x_1 + 5x_2 & \geq 45 \\
\text{(B)} & x_1 + x_2 & \geq 6 \\
\text{(C)} & 8x_1 + 5x_2 & \leq 0 \;\; (\textit{soft}) \\
& x_1, x_2 & \in \{1, 2, \ldots, 5\}.
\end{array}
\tag{3.12}
$$

One easily verifies that the OIP (3.12) is confined and can thus be reduced to an integer linear program with minimization objective $8x_1 + 5x_2$. Figure 3.7 shows the feasible region of the LP relaxation shaded in gray, and delimited by the constraints (the feasible region of soft constraint (C) is shaded light gray). The black circles are the points that are feasible to the integer problem, the white circles are the points that are integer but are outside the feasible region defined by the (hard) problem constraints. Also plotted is one *isocost line* (arbitrarily starting at point $(3,5)$) and denoted by the equation $z = 8x_1 + 5x_2$. All isocost lines are parallel to the plotted one, and the intersection of the feasible region with the leftmost possible isocost line (minimizing z) yields the LP optimum.

To illustrate the search process of WSAT(OIP) and several improvements by this example, we will consider in order (a) the search process with vanilla parameters, (b) search with tabu tenure $t = 1$, (c) search with constraint weights λ, (d) initialization by rounding of the LP optimum, and (e) search after constraint LP rebounding.

The Figures 3.8 illustrate the moves of WSAT(OIP) to solve the problem (3.12) using different strategies. First, Figure 3.8(a) depicts the progress of vanilla WSAT(OIP) after an initialization to $(0,0)$. Each diamond plots one variable assignment visited during the search. Each arrow plots the trajectory between two assignments, and each arrow is marked by the constraint (A,B, or C) that is selected for partial repair. If different selected constraints lead to the same move, the transition arrow is marked with more than one letter (e. g. A,B).[9]

To eliminate the random aspect from this discussion, we do not allow random moves, i. e. we always start at point $(0,0)$ (setting $p_{zero} = 1$), always take the best possible move ($p_{noise} = 0$) and always pick a hard constraint

---

[9] In the particular example we are lucky in that no branching trajectories exist.

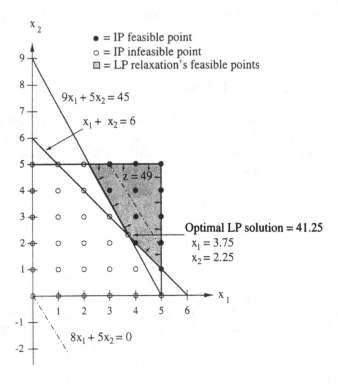

Figure 3.7. Graphical Interpretation of problem (3.12).

for repair when one is violated ($p_{hard} = 1$). Otherwise, we employ standard parameter settings. Thus, **Maxstep** is 2, which means that individual variable changes are allowed up to two units in this example, leading to arrows of up to length 2.

The vanilla search 3.8(a) starts by greedy moves along the x_1 axis, since the score improvement is larger along x_1 than x_2. For another 3 moves, WSAT(OIP) continues moving towards the feasible region, first reaching A's feasible region at $(5,0)$, then the overall feasible region and then reaching the local optimum $(5,1)$ with an objective function value of 45. Now, both hard constraints are satisfied and the soft constraint C is selected for repair. With respect to the score gradient, decreasing x_2 appears to be the best alternative, reaching the previously visited point $(5,0)$. At this point, the search starts cycling because the noise level is zero.

Tabu Search. The tabu search 3.8(b) proceeds similarly, except that the transition back to $(5,0)$ is tabu and the search happens to enter the feasible region

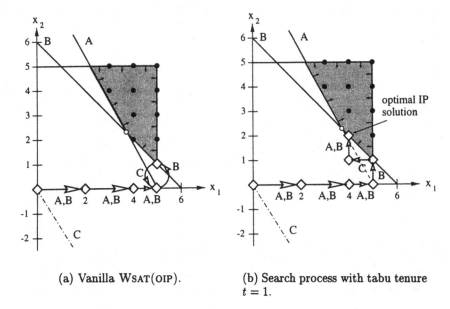

(a) Vanilla WSAT(OIP).

(b) Search process with tabu tenure $t = 1$.

Figure 3.8. Search trajectories of different WSAT(OIP) strategies.

at the IP optimal point (4,2) with an objective function value of 42. We notice that although the tabu element prevents cycling in this case, the search reaches the optimal solution only after a detour. The tabu mechanism was also critical to find good solutions in several case studies.

Observation of 3.8(a) reveals that the mistake of the vanilla search was that it did not turn at (4,0). Instead of moving more directly towards the IP optimum, the score suggested that the better move was to achieve feasibility of constraint A. Using this example, we will discuss more elaborate techniques to alleviate the situation. Some of the following techniques have been applied within the application case studies.

Constraint Weights. In the vanilla search at point (4,0), why did the score suggest to move along x_1? Clearly, in terms of feasibility, (4,2) is superior as it is feasible with respect to both constraints A *and* B. Nevertheless, the violated of soft constraint C dominates and the score indicated to take the move (5,0). In this example, however, moving towards the feasible region first would have yielded a better solution. One way to achieve this with WSAT(OIP) is to increase the weights of hard constraints. The trajectory in Figure 3.8(c) is the result of setting λ to a large integer for A and B, say 100 (illustrated by bold lines). Constraint weights were also critical in the case study on capacitated production planning.

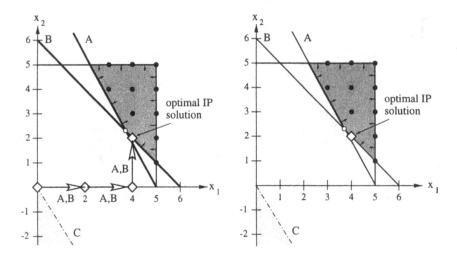

(c) Search process with weights on hard constraints.

(d) Search process initialized by rounding the LP optimal solution ∘.

Figure 3.8. (cont) Search trajectories of different WSAT(OIP) strategies.

Initialization by Rounding LP Solutions. Sometimes it is useful to start from an initial solution which is close to the LP optimum, as sketched in the previous section. Figure 3.8(d) depicts this search process which consists of no moves at all, as the initial solution is already IP optimal (which is of course rarely the case in practice). Notice that optimality is proved here by the LP optimum of 41.25: there exists no integer solution with a cost better than 42 and above the LP optimum.

Constraint Rebounding. In section 3.1.3, we mentioned that it can be advantageous to relax the threshold of a soft constraint and enlarge its feasible region. As we will see, our example is such a case. By solving the LP relaxation, we determine that the soft constraint C can never be fully satisfied as the LP optimum is 41.25. Relaxing the right-hand-side of inequality C to a valid lower bound is a confined rebounding and yields the system (3.13), which has the same set of solutions as the original problem (by proposition 3.1.3).[10]

---

[10] Notice that in general there is more than one soft constraint and more complex rebounding techniques need to be applied if the bounds are not inherently provided in the model.

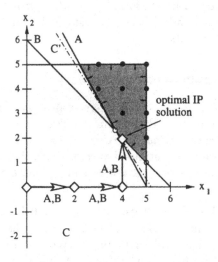

(e) Search process after constraint
LP rebounding.

Figure 3.8. (cont) Search trajectories of different WSAT(OIP) strategies.

$$
\begin{array}{lll}
\text{(A)} & 9x_1 + 5x_2 & \geq 45 \\
\text{(B)} & x_1 + x_2 & \geq 6 \\
\text{(C')} & 8x_1 + 5x_2 & \leq 41.25 \; (= Z_{\text{LP}}) \; (\textit{soft}) \\
& x_1, x_2 & \in \{1, 2, \ldots, 5\}.
\end{array}
\qquad (3.13)
$$

The search process on the rebounded system is illustrated in figure 3.8(e). Through the rebounding, the assignments along the trajectory are now all in the feasible region for C', and the search concentrates on entering the feasible region of the hard constraints. Notice that after the rebounding, the objective function values are shifted down by $41.25 - 0$, thus the optimal IP solution is now 0.75.

3.5 Related Work

There are three principal lines of related work. (i) General-purpose heuristics for integer programming, (ii) local search strategies for constraint satisfaction (CSP) problems, and (iii) domain-specific heuristics for problem classes like set-covering, generalized assignment, or time-tabling. We will focus on (i) and (ii) and will not discuss the numerous domain-specific heuristics (iii) here.

In the following, we will present a thorough review of the literature on general-purpose heuristics and summarize the problem classes used for experimentation. As one indicator of problem difficulty, the largest problem sizes are additionally reported in the form $^{(n \times m)}$, where n is the largest number of variables ('columns') and m is the largest number of constraints ('rows') reported in the experiments.

3.5.1 Integer Programming Heuristics

We start by discussing related work on general purpose heuristics for integer programming from the OR literature. There are two principal characteristics of *integer programming heuristics*. They either relax the integrality constraints and operate on continuous variables. Or, they adhere to the integrality constraints and perform local moves in the space of integer solutions, like WSAT(OIP) does.

Integrality Relaxing Heuristics. A pioneering 0-1 integer programming heuristic which combines a variety of techniques is the *pivot&complement heuristic* (P&C) of Balas and Martin [9]. It exploits the fact that an optimal solution to a 0-1 IP problem can be found at one of the extreme points of the linear programming feasible region.

P&C has both of the above characteristics and operates in two phases. The first ("search") phase performs mainly pivot moves (in the simplex tableau) attempting to find at a good feasible 0-1 solution. Once feasibility has been achieved, the second ("improvement") phase attempts to improve the solution objective by flipping in turn one variable, a tuple or a triplet of variables. Experimental results with P&C have been reported in [9] for capital budgeting $^{(200 \times 30)}$, set covering $^{(905 \times 200)}$, set partitioning $^{(14 \times 88)}$ (crew scheduling). Tabu search enhancements of P&C (treating P&C as a black-box subroutine) a have been evaluated on multi-constrained knapsack problems $^{(105 \times 30)}$ [3, 104] as well as special set covering $^{(45 \times 331)}$ and miscellaneous problems $^{(224 \times 201)}$ [3].

Another strategy that superimposes tabu search principles to extreme-point transitions has been given by Løkketangen and Glover [103]. It uses advanced tabu search strategies for diversification and a learning approach called 'target analysis'. Experimental results are reported for multi-constrained knapsack problems $^{(90 \times 30)}$. Recently, Glover and Laguna [59, 60] have proposed a theoretical basis for IP heuristics. Their approach shares a foundation with a framework for generating cutting planes for IP problems, but has not yet been experimentally evaluated.

Integer Local Search. The next of kin of WSAT(OIP) is the class of integer local search heuristics. Among the few strategies which have been reported is a simulated annealing strategy (GPSIMAN) by Connolly [33] as well as a refinement (RFSA) and a variation (PISA) thereof by Abramson *et al.* [4]. RFSA adds the search space reduction technique of Section 3.3.3 to GPSIMAN, and PISA uses a different (inferior) neighborhood transition approach.

In essence, GPSIMAN proceeds as follows. In each iteration, suppose we start from a feasible assignment A. A random 0-1 variable is selected and flipped. Then, an attempt is made to restore feasibility of the changed assignment, yielding A'. Next, the score $C(A)$ is compared to $C(A')$ and the move is always accepted provided it is not deteriorating the score (i. e. $\Delta C = C(A') - C(A) \leq 0$). If A' is worse than A, the move is accepted with a probability $e^{\Delta C/T}$, where T is a *temperature* that is decreased according to an annealing schedule. The annealing schedule starts from a high temperature (many deteriorating moves allowed) and is slowly reduced. Occasionally, a temporary reheating phase takes place.

The feasibility restoration of GPSIMAN and RFSA ('restore feasibility') proceeds by ranking the variables according to a 'help-score' and choosing the best variable. Unlike WSAT(OIP), the computation of the help-score does not predict the true effect of flipping a variable. Instead, it computes an intricate measure of the repair effect, taking into account a 'criticality' of the violated constraints. Details of the computation can be found in [4] or [33]. In contrast, PISA ('penalize infeasibility') takes a different approach by incorporating violated constraints into the objective function, and not restoring feasibility after each flip. Similarly as WSAT(OIP), PISA allows the search to move through infeasible regions of the search space. Abramson *et al.* do not commit to a particular penalty function but find PISA to have inferior performance than RFSA. In [4], a number of disadvantages of this scheme are discussed that do not apply to WSAT(OIP), e. g. "a new cost function is required every time a new problem is encountered." The version of GPSIMAN we had available is implemented in Fortran.

WSAT(OIP) is related to both PISA and RFSA: Although WSAT(OIP) can move through infeasible regions, its repair strategy (p_{hard}) may be set to immediate feasibility restoration as well. We see the principal differences to GPSIMAN in the score computation of WSAT(OIP), the applied principles from tabu search, and the fact that WSAT(OIP) operates on OIPs according to the Walksat-Principle.

Experimental Comparison with GPSIMAN. We have evaluated the GPSIMAN solver on our benchmarks (RFSA has ceased service and was not available). However, with exception to the radar surveillance problems (Section 6.1), it did not succeed in finding acceptable solutions. For the radar surveillance problems, the results were not competitive to those of WSAT(OIP), despite GPSIMAN was allowed more than two orders of magnitude the runtime of WSAT(OIP). The results are consistent with previously reported experiments by Abramson *et al.* [4].

For the course assignment problems, GPSIMAN was not able to find satisfactory solutions to even small instances.[11] Also, the timetabling problems in Chapter 5 (all tightness levels) were beyond the limitations of GPSIMAN, which gives supporting evidence for Abramson *et al.*'s conclusion that "although the RFSA approach performs better than the PISA approach, it fails for problems that are heavily constrained". GPSIMAN does not support gen-

---

[11] Unfortunately, the implementation produced illegal (super-optimal) solutions when given more promising parameters.

eral integer variables and therefore not applicable to the Chapter 7 problems. In [33] GPSIMAN experiments are described on quadratic assignment $^{(506 \times 458)}$, grids-and-crosses $^{(256 \times 78)}$, knapsack $^{(100 \times 1)}$, processor-communication $^{(66 \times 698)}$, graph coloring $^{(248 \times 630)}$, and class timetabling $^{(468 \times 492)}$. RFSA and PISA have been evaluated and compared on large set partitioning problems $^{(78186 \times 492)}$ using extensive preprocessing and dynamic search space reduction [4].

A Simulated Annealing Code. A recent approach, INTSA, by Abramson and Randall [5] combines different neighborhood transition schemes of simulated annealing for different combinatorial problems. INTSA is reported to perform well in comparison with GPSIMAN and a IP branch-and-bound solver (OSL). The goal of the framework is to automatically choose an appropriate SA neighborhood based on the given algebraic problem description. For example, choose value changes for graph coloring, but choose a 2-opt move for the type of constraint used to represent traveling salesman problems.

While this is clearly an appealing idea, it remains to be shown that INTSA will be able to handle problems with mixed types of constraints (in addition to the pure problems handled by problem-specific SA implementations). In fact, as [5] note, "the INTSA results were actually gathered from a number of different programs, each of which handled one of the classes of constraints rather than one program which could differentiate the constraint class and choose the appropriate algorithm." INTSA has been evaluated on quadratic assignment (30 facilities), traveling salesman (666 cities), graph coloring (300 nodes, 740 edges), bin packing (500 items, 201 bins), and generalized assignment (8 agents, 32 jobs).

We would very much like to see an approach that combines the strengths of INTSA and WSAT(OIP): different neighborhood schemes combined with flexibility in the type of constraints. One possible scenario would be to employ the Walksat Principle for dispatching the repairs to be made.

CSP Tabu Search as a General Problem Solver. Very recently, an independent approach of iterative repair to linear integer constraints was published by Nonobe and Ibaraki [116] (previously presented at APORS-97 in Melbourne, December 97; WSAT($\mathcal{PB}$) appeared at AAAI-97). It shares with this work the iterative repair approach to ILP problems, and also provides a comprehensive empirical study with encouraging results. Technically, the approaches are less similar. First, the framework by Nonobe and Ibaraki does not start from SAT local search, and hence uses different local moves ('shift' and 'swap') instead of the atomic flip moves performed by WSAT(OIP). Further, to approach optimization problems, [116] introduces a mechanism to tighten a bound on the objective function and an additional control mechanism. This is in contrast to our use of OIP for this purpose. Further, [116] employs an open definition of CSP as base representation and hence does not address combinations with linear programming. Also, [116] employs solely 0-1 variables, and while two of the experimental studies are on similar problem types, no application to tight 0-1 ILP feasibility problems seems to be provided in [116]. It would be interesting to further compare the two approaches.

3.5.2 Local Search in Constraint Satisfaction

The other line of work on domain-independent local search has taken place in the context of constraint satisfaction problems (CSPs) in artificial intelligence. For the purpose of the discussion, we distinguish between binary CSPs (in which all constraints involve exactly two variables) and non-binary CSPs. Also, we distinguish between an extensional representation (in which constraints are represented by an explicit set of allowed or forbidden variable-value tuples) and intensional representations. Of course, algebraic constraints are non-extensional and generally non-binary representations.

One of the earliest approaches of heuristic search in constraint satisfaction is the *min-conflicts* heuristic by Minton *et al.* [109] (previously published 1990). It has been formulated both as a backtracking algorithm and as a hill climbing strategy. The basic principle is the same as in SAT local search, namely to "select a variable that is in conflict, and assign it a value that minimizes the number of conflicts" [109]. Min-conflicts was evaluated on graph coloring (binary), the n-queens problem (binary), and scheduling of the Hubble Space Telescope (non-binary).

Different methods in the same spirit were evaluated for graph coloring and frequency assignment by Hao and Dorne [68]. The min-conflicts strategy was enriched with noise and applied to randomly generated binary MAX-CSP problems by Wallace and Freuder [147].

Another line of work is the connectionist approach GENET by Wang, Tsang, Davenport *et al.* [38, 39]. GENET is an iterative repair network approach that operates by a heuristic learning rule. Although GENET was initially formulated for binary extensionally represented CSPs, various extensions to non-binary non-extensional CSPs have taken place, among them the work on E-GENET [99, 100]. In E-GENET, more expressive constraints have been studied (each on one type of benchmark): A 'linear-arithmetic' constraint (cryptarithmetic puzzles), an 'atMost' constraint (car sequencing), a 'disjunctive' constraint (Hamiltonian path), and a 'cumulative' constraint (simple scheduling problem).

3.6 Summary

In this chapter, we have introduced WSAT(OIP), an integer local search method which operates on an algebraic problem representation. WSAT(OIP) generalizes Walksat, a successful local search procedure for propositional satisfiability (SAT), to more expressive constraint systems.

For the purpose of the generalization, we have introduced over-constrained integer programs (OIPs), a constraint class which is closely related to integer programs. OIP allows for a natural generalization of the principles of SAT local search to integer optimization. Further, it has been shown that OIPs are a special case of integer linear programs and permit combinations with linear programming for bound computation, initialization by rounding, search space reduction, and feasibility testing. The representation is similar enough to integer programs to make use of existing algebraic modeling languages as front-end to a local search solver. To improve performance on realistic problems, WSAT(OIP) incorporates strategies from Tabu Search. On a graphic example, we have illustrated different sub-strategies for local moves in WSAT(OIP), i.e. a tabu element, constraint weights, initialization by rounding and constraint rebounding.

4. Case Studies Methodology

"Any choice of [benchmark] problems is open to the criticism that it is unrepresentative. There is another way, however. One can investigate how algorithmic performance depends on problem characteristics. The issue of problem choice, therefore, becomes one of experimental design. Rather than agonize over whether a problem set is representative of practice, one picks problems that vary along one or more parameters."

John N. Hooker in [75]

The previous chapter has described new methods for integer local search and presented the WSAT(OIP) procedure. The next three chapters will empirically investigate the performance of WSAT(OIP) in a number of realistic case studies. In between, this chapter reflects on issues and goals of our experimental analysis.

There are three chapters for three different aspects of WSAT(OIP): Chapter 5 investigates WSAT(OIP)'s ability to solve difficult 0-1 integer feasibility problems. Chapter 6 concentrates on 0-1 integer optimization problems. Finally, Chapter 7 focuses on the extension of WSAT(OIP) to (non 0-1) finite domain problems.

There can be different goals of experimental analysis [124, 75, 11]. Because integer local search is at an early stage, our case studies mainly investigate two central aspects: Domain-independence and applicability to optimization in practice. To demonstrate domain-independence, we draw applications from a range of integer optimization problems. To support the claim of practicality, we highlight the aspects that we believe matter for practical concerns. This chapter contemplates criteria of success for practical optimization methods and motivates the case studies on the grounds of those criteria.

4.1 Optimization in Practice: Criteria of Success

Theoretical analysis of algorithms is usually concerned with aspects of *worst-case or average case resource requirements*. Normally, one is interested in bounds on resource usage such as time or memory, usually under varied problem parameters such as size. Additionally, for approximation algorithms, one attempts to derive lower bounds on the quality of the solutions. However, at present most practical optimization algorithms for *NP*-hard problems are beyond the scope of rigorous theoretical analysis [75, 86, 11], even worse so when applied to realistic problem classes. In this situation, one needs to resort to experimental testing.

When moving from theoretical to experimental analysis, one is faced with an unfamiliar amount of freedom. Factors that often limit theoretical analyses disappear, such as restrictive assumptions on the instance distribution or restrictions on the algorithmic properties to investigate. Hence, careful decisions need to be made what aspects an empirical evaluation should investigate. The central decisions are how to define algorithmic performance and to select a set of 'typical' problems to evaluate.

Normally, given a problem instance, performance is measured in terms of (i) *time to obtain the first or best solution*, or (ii) *best solution quality obtained in limited time*. Orthogonally, however, it is critical to investigate the *variation* of these measures on a given instance distribution. We refer to performance variation over a given instance distribution as *robustness*. Note that there is an another issue of robustness, namely performance variation when *changing* the instance distribution (i. e. considering different problems), which we call *flexibility*.

On first sight, robustness might be regarded as a question of secondary importance in comparison to solution quality or runtime. But in fact, it is of critical importance and inseparable. In particular for *NP*-hard problems, an algorithm that performs well on one set of problem instances is hardly of any practical use if small variations of the instance parameters break it (if its runtime changes from 4.83 seconds, precisely measured on one given instance, to 2 weeks on the next).

There are several aspects of robustness. First, we are interested in the *scaling* of runtime with *increasing problem size* (keeping other problem characteristics similar). The second aspect is scaling of runtime with *increasing constrainedness*, i. e. how does performance vary as additional constraints are thrown at the problem.[1] A third aspect is *residual robustness*, i. e. robustness under minor variations of instance characteristics which only remotely affect size or constrainedness.

4.1.1 Scaling with Increasing Problem Size

At the center of most theoretical algorithm analysis is the question how an algorithm's performance varies with increasing problem size. Despite the importance commonly attributed to this question on a theoretical level, it is sometimes neglected in experimental studies.

Sometimes, there are good reasons not to study scaling with size, especially on real problem instances, where one usually has no handle on the size

---

[1] We will not attempt to give a rigorous definition of 'constrainedness'.

of the problem. Even if instance size varies, real problems sometimes happen to vary strongly along a number of characteristics and size may appear as 'just another parameter'. It is then difficult to isolate size from the other parameters. However, if artificially generated problems are studied, there is normally no reason not to investigate scaling (care must be taken not to change the problem characteristics when crafting instances of different size). We measure problem size by the number of variables and constraints of a given encoding. The scaling behavior is of practical importance since real problems are often large—typically at least as large as state-of-the-art technology can handle.

Moreover, what makes scaling critical is that different algorithms exhibit *different* scaling behavior. Empirically, what is a excellent algorithm for small problems may not be applicable to large-scale problems. Conversely, a heuristic that works well for large problems may not have the desired properties for small problems (i. e. because it is approximate and one might care for optimality). Scaling of local search has previously been examined on hard randomly generated satisfiability problems and sub-exponential (average) scaling was observed [53, 118]. In our case studies, we examine the scaling behavior of integer local search for realistically structured (randomly generated) covering problems as well as for real course assignment problems in Chapter 6.

4.1.2 Scaling with Increasing Constrainedness

A second dimension in scaling occurs with increasing constrainedness. Investigating algorithmic behavior along this dimension is important in particular as the typical practitioner's approach to constraint problems is iterative repair: "State some known constraints and find a solution by invoking a solver. Observe that it exhibits certain unliked characteristics and state additional constraints that disallow them. Re-solve and iterate." In this typical spiral process, it is critical that both the loosely constrained and the more tightly constrained problem can be solved.

In recent years, there has been some interest in AI in studying the algorithmic performance across different degrees of constrainedness. For several problem domains and algorithms, an easy-hard-easy pattern has been observed in time-to-first-solution as the problem constrainedness is being increased. Most of these studies have investigated performance on randomly generated problem instances. As yet, there is no generally accepted way to quantify the constrainedness of a problem instance, although measures for the constrainedness of an ensemble have been given [51, 110, 151].

Chapters 5 (timetabling) and 7 (production planning) shed some light on the issue of scaling with increasing constrainedness on real problems. While in the first study we use the number of solutions as a rough measure of constrainedness, the latter uses a parameter of the input problem, i. e. resource capacity.

4.1.3 Flexibility and Residual Robustness

Another requirement of practical optimization methods is *flexibility*. When a method is specifically tailored to a narrow problem class (e. g. set covering *or* generalized assignment *or* time-tabling etc.), incorporating additional constraints to solve a closely related problem usually requires adjusting the algorithms or replacing the strategies altogether. Flexibility is investigated both across the different case studies (if a method is domain-independent, it will have to be flexible) and within the case studies (each of the considered problems includes a variety of different constraints).

The last aspect of runtime variation is *residual robustness*: After factoring out issues of size and constrainedness, we are left with a residuum of performance variation on a given instance distribution. For real problems, however, it may be difficult to obtain several instances with similar parameters. One solution is to perturb the input parameters of a given instance, thereby generating 'pseudo-real' problems.[2] We refrained from perturbing real problems to avoid the difficulty of choosing which factors to perturb. In most cases, the case studies examine several similar instances of a given problem.

4.2 The Problem Class Selection

In recent years, the use of randomly generated benchmark problems has increasingly been criticized for empirical evaluation of algorithms, e. g. [88]. To address the need for a realistic assessment of optimization technology, our choice of benchmark problems focuses on problems 'as real-world as available'. All benchmark problems have either been studied in the recent AI or OR literature (timetabling and sports scheduling), are the result of industrial cooperations (radar surveillance covering and capacitated production planning) or originate from operating applications (course assignment, sports scheduling).

The benchmarks have been selected to examine the four requirements for practical optimization methods stated above: (i) scaling with problem size and (ii) constrainedness, (iii) flexibility and (iv) residual robustness. All benchmark problems under consideration share that they stem from NP-hard problem classes, have a large number of variables and constraints, contain a heterogenous set of constraints, and are difficult for the best general state-of-the-art optimization techniques available (both scientific and commercial packages). The set of benchmarks diverges from the classical pure problems (e. g. set covering, set partitioning, generalized assignment etc.) and involves complicating side constraints which, in most cases, would have prevented a direct application of domain-specific heuristics from the literature. All case studies are on integer *linear* problems. All but one case study are on 0-1 integer optimization (binary variables).

---

[2] A term used by Toby Walsh in personal communication.

Two case studies (Chapter 5) address the ability to solve hard feasibility problems, a property not commonly addressed by optimization benchmarks. Such problems are typically difficult for general-purpose integer optimization methods like IP branch-and-bound. We did not investigate graph coloring problems because several studies of iterative repair exist for this domain [135, 68, 109]. Clearly, integer optimization problems vary largely and the particular selection only covers a small fraction. The claim on domain-independence should hence be viewed in relation to the state-of-the-art ILP solver technology.

The benchmark problems have the following size characteristics (maximal number of variables and constraints): Radar surveillance covering (10989×14595), course assignment (8404×11350), the Progressive Party Problem (4632×30965), ACC Basketball (1339×3053), and production planning (using finite domain variables, 7520×3047). Although problem size does not directly imply hardness, it is a relevant problem characteristic and we notice that the size of the benchmarks is larger than in many previous studies of general-purpose methods.

With exception to capacitated production planning (which contains proprietary data), all benchmark problems have been made available through the Constraints Archive at http://www.cirl.uoregon.edu/constraints/.

4.3 The Empirical Comparisons

To demonstrate the performance and range of applicability of integer local search, this study takes a competitive approach to performance evaluation. It focuses on comparing WSAT(OIP) to other *general-purpose* optimization frameworks, which have been described in Chapter 2. Whenever appropriate, we thus compare to IP/MIP branch-and-bound (CPLEX 5.0 [80]), constraint programming solvers (Oz [137] and ILOG Solver [79] approaches from the literature), and GPSIMAN, a domain-independent simulated-annealing heuristic [33, 4].[3] The employed constraint programming approaches incorporate some domain knowledge in the form of enumeration heuristics or suitable problem factorizations.

In order to compare the results of the WSAT(OIP) heuristic to exact algorithms, the exact methods are run in "heuristic mode" [9], i.e. the best solution found within a given time limit is reported (occasionally, second-best solutions are also reported if qualities are similar but time differs significantly), if optimality can be proved, this is reported. To evaluate the experimental results in absence of provably optimal solutions, we employ methods for generating valid lower bounds, i.e. linear relaxation and Lagrangean relaxation.

---

[3] I thank David Abramson and Marcus Randall for providing the GPSIMAN solver of David Connolly and for useful suggestions to parameter settings.

For the MIP branch-and-bound experiments, the CPLEX 5.0 MIP optimizer [80] is used as it is commonly regarded as one of the fastest general-
purpose MIP-optimizers and has been in commercial use for over 10 years.
CPLEX 5.0 utilizes state-of-the-art algorithms and techniques, including cuts
(cliques & covers), heuristics, a variety of branching and node selection strategies, and a sophisticated mixed integer pre-processing system [80]. We often
run CPLEX with *standard parameters*, which involves *automatic* control of
several MIP subroutines (such as heuristics, branching, cut generation). The
standard parameter settings are usually non-trivial to improve upon (in many
cases we report on non-standard settings as well). It should be recognized that
CPLEX is the product of several man-decades of development and research,
whereas the WSAT(OIP) implementation is comparatively simple.

Mostly, run-times are reported but no memory requirements even though
in some applications space usage may be an important issue. We neglect
the issue in our study, but point out that the memory usage may differ
significantly for the different frameworks: Tree-search approaches like CPLEX
and CP sometimes occupy hundreds of megabytes of main memory while
integer local search uses constant memory during the search. To compensate
for exaggerated memory usage, all runtimes are measured as wall clock time,
which purposefully incorporates a penalty for paging.

5. Time-Tabling and Sports Scheduling

This chapter investigates two difficult time-tabling/scheduling problems, 'the Progressive Party Problem' (PPP) and scheduling of the Atlantic Coast (basketball) Competition of 1997/98 (ACC). Both problems were recently introduced and solved in the literature [115, 136]. The previous results demonstrate that finding feasible solutions for these problem is challenging, even when using approaches that incorporate domain-knowledge.

Both problems can be encoded with 0-1 integer constraints (a model for PPP has been given in [136] and a model for the full ACC problem will be presented here), but no approaches have previously been reported to find solutions directly from 0-1 encodings of these problems.

This chapter reports on experiments of integer local search given 0-1 integer models of problem instances of PPP and ACC. For both problems, the experimental results will be compared to the previously reported results. Moreover, we will study the performance of local search with increasing problem constrainedness. In particular, for the ACC problem, an extensive study of integer local will be presented that investigates the scaling of runtime with increasing problem constrainedness.

5.1 The Progressive Party Problem

The problem in the first case study, "the progressive party problem", was recently introduced in a comparison between constraint programming and integer linear programming [136]. A main result of the study is that the problem appears to be beyond the size limitations of integer linear programming (ILP) but can be solved using constraint propagation and chronological backtracking. Our experiments show that the problem can be solved significantly faster using WSAT(OIP). Further, we look at slight variations of the instance given in [136] and find that local search is robust with respect to the modifications. On the other hand we were not able to find a constraint program that could solve all of our test problems. To solve the problem with WSAT(OIP), we factor it into two stages. In the first stage, a small number of principal variables are explicitly enumerated (e. g. using constraint programming), while in the second stage, the variables valued in stage one are

Table 5.1. Boat specifications. The entries are boat number i, spare capacity $K_i - c_i$ and crew size c_i.

boat	cap	crew	boat	cap	crew	boat	cap	crew
1	6	2	15	8	3	29	6	2
2	8	2	16	12	6	30	6	4
3	12	2	17	8	2	31	6	2
4	12	2	18	8	2	32	6	2
5	12	4	19	8	4	33	6	2
6	12	4	20	8	2	34	6	2
7	12	4	21	8	4	35	6	2
8	10	1	22	8	5	36	6	2
9	10	2	23	7	4	37	6	4
10	10	2	24	7	4	38	6	5
11	10	2	25	7	2	39	9	7
12	10	3	26	7	2	40	0	2
13	8	4	27	7	4	41	0	3
14	8	2	28	7	5	42	0	4

propagated through the theory, and the remaining subproblem is attacked with local search.

5.1.1 Problem Description and Formulation

In the integer local search approach, we employ a 0-1 model similar to the one used by Smith *et al.* [136]. The problem model is large and incorporates a variety of different constraints which suggested that it would be an interesting test case for integer local search.

The problem scenario is an evening party in the context of a yachting rally. Certain boats are selected to be hosts, and the crews of the remaining boats in turn visit the host boats for several successive half-hour periods. The crew of a host boat remains on board to act as hosts while the crew of a guest boat together visits several hosts. Every boat can only host a limited number of guests at a time and crew sizes are different. Table 5.1 reports boat capacities and crew sizes. There are six time periods. A guest boat cannot revisit a host and guest crews cannot meet more than once. The problem facing the rally organizer is that of minimizing the number of host boats (presumably for reasons of supply logistics): Certain boats are constrained to be hosts, and selecting the hosts among the remaining boats is stated as part of the problem.

We do not claim that this problem is of immediate practical significance; however, it has the advantage of being a well-studied hard time-tabling problem with a variety of constraints. The variables in the problem are the following: $\delta_i = 1$ iff boat i is used as host boat. Variables $\gamma_{ikt} = 1$ iff boat k is a guest of boat i in period t. Constant c_i is the crew size of boat i and K_i is its total capacity. The objective is to minimize the number of hosts $\sum_i \delta_i$, subject to:

Constraints CD. A boat can only be visited if it is a host boat.

$$\gamma_{ikt} - \delta_i \leq 0 \text{ for all } i, k, t; \ i \neq k.$$

Constraints CCAP. The capacity of a host boat cannot be exceeded.

$$\sum_{k, k \neq i} c_k \gamma_{ikt} \leq K_i - c_i \text{ for all } i, t.$$

Constraints GA. Each crew must always have a host or be a host.

$$\delta_k + \sum_{i, i \neq k} \gamma_{ikt} = 1 \text{ for all } k, t.$$

Constraints GB. A guest crew cannot visit a host boat more than once.

$$\sum_t \gamma_{ikt} \leq 1 \text{ for all } i, k; \ i \neq k.$$

An additional set of 0-1 variables was introduced to state the meet-once restrictions. $m_{klt} = 1$ if boats k and l meet at time t. This simplifies the ILP model described in [136].[1]

Constraints U. Link m_{klt} with γ_{ikt}.

$$\gamma_{ikt} + \gamma_{ilt} - m_{klt} \leq 1 \text{ for all } k, l, t; \ k < l.$$

Constraints M. Every pair of hosts can meet at most once.

$$\sum_t m_{klt} \leq 1 \text{ for all } k, l; \ k < l.$$

With B boats and T time periods, the problem has $O(B^2 T)$ variables and $O(B^2 T)$ constraints in this formulation. Smith *et al.* note that the CP representation is more compact and has "far fewer constraints and variables than the ILP". This is not the case since the number of both constraints and variables is actually $O(B^2 T)$ in both encodings (even in the improved ILP model in [136]).

Although the problem is formulated as an optimization problem, given the particular description of the participating boats the task is to find a

---

[1] The original ILP description [136] is $m_{klt} = 1$ **iff** boats k and l meet at time t. The modification simplifies the problem and saves approximately 30K clauses. According to Sally Brailsford (personal communication) this had been tried in the ILP model.

feasible assignment with 13 host boats. Every solution with 13 hosts is optimal because the capacity constraints cannot be met with 12 hosts even for a single time period. Solving the problem can be divided into two stages: (i) selection of the host boats, and (ii) assignment of guest boats to hosts for all time periods. It turns out that the *spare capacity* of the boats is a good indicator of whether a boat should be host or guest, so after forcing special boats to be hosts (e.g. the rally organizer), the remaining hosts were selected by decreasing spare capacity (the spare capacity of a boat is its total capacity minus its crew size). In both the ILP and the CP approach, Smith *et al.* treat both stages of the problem. However, the search-space for a particular host selection is too large to be explored exhaustively within hours of computation. This shows that solving stage (ii) by itself is a hard subproblem and we therefore focus on stage (ii): Finding a guest allocation given a fixed selection of hosts. Thereafter we will outline a strategy that captures both stages.

Smith *et al.* report the problem could not be solved with a commercial integer programming tool (XPRESSMP, using a variety of tricks) because it appears to be beyond the size limitations of ILP.

5.1.2 Experimental Results and Comparison

For the experiments, we use the original problem instance of Smith *et al.* and randomly vary the host selection to produce 5 additional instances. For all instances, we keep the original description of boat capacities and crew sizes. After fixing the 13 hosts and performing constraint propagation as an efficient preprocessing, the original problem has 4632 variables and 30965 remaining clauses in pseudo-Boolean formulation. WSAT(OIP) finds a feasible guest allocation in 5.5 seconds (averaged over 20 successful runs on a SPARCstation 20) using a tabu memory of size 1 and initializing with a bias of $p_z = 0.9$. Additionally, setting up the constraints from an abstract representation requires around 15 seconds. Table 5.2 summarizes the results.

For comparison, Smith *et al.* report 27 minutes of runtime of their ILOG Solver program on a SPARCstation IPX. To reproduce the results, we implemented the described modeling in Oz,[2] a concurrent constraint language [137].[3] We used constraints and a labeling strategy similar to the one described by Smith *et al.*. Although our constraint program was able to solve the original instance in 8 minutes (on a SPARCstation 20), we could not find a labeling strategy that was able to solve all sample instances.[4]

It has been claimed [136] that the progressive party problem can be solved with constraint programming in a straightforward way. Our experiments confirm this for the original problem instance, but we find that slight variations

---

[2] I thank Jörg Würtz and Thorsten Ölgart for modeling the progressive party problem in Oz.

[3] Publically available from http::/www.ps.uni-sb.de/oz/.

[4] Jörg Würtz, personal communication.

Table 5.2. Experimental results for variations of the Progressive Party Problem. The columns are: Selected hosts, total sum of host spare capacities h, total sum of guest crew sizes g; percentage of total capacity used as a measure of constrainedness (%cap $= g/h$). Runtimes averaged over 20 runs of WSAT(OIP), Maxmoves$= \infty$, flip-rate 1.1 K-flips/s.

host boats	h	g	%cap	WSAT(OIP)
1–12,16	100	92	.92	2.9s
1–13 (orig)	98	94	.96	5.5s
1,3-13,19	96	92	.96	6.4s
3-13,25,26	98	94	.96	8.8s
1–11,19,21	95	93	.98	31.6s
1–9,16–19	93	91	.98	42.5s

can make the problem too difficult to solve in hours of computation.[5] Because CP is not a particular algorithm but subsumes a wide variety of techniques to operationalize constraint solving, no general conclusion can be drawn about its performance on the particular problem. Stronger propagation, better labeling, randomization [63] or search strategies might be able to improve the performance robustness on this problem.

The authors of [136] also report an integer programming approach (using XPRESSMP) given the stage (ii) problem, in which a problem with up to 15 boats and 4 time periods could be solved.

Embedding into constraint programming. To solve both stages of the problem, we propose a loose coupling of systematic and local search. The approach simply enumerates the principal variables heuristically (in this case the δ_i's, stage (i)), then performs constraint propagation/simplification and applies local search to solve the remaining subproblem (stage (ii)). In our implementation, we use an embedding of WSAT(OIP) into the constraint language Oz. The advantage of using a constraint language is the high-level support for problem modeling and solution checking. Oz additionally offers the use of computation spaces which simplifies the embedding of a solver like WSAT(OIP) into CP.

Notes on the Experiments. Before using the two-stage approach, we experimented with local search on a version of the problem that included host selection. Observation of the local search process revealed that host selection and guest allocation were mixed and the host selection was changed almost as often as the guest allocation, which seemed to be an unreasonable strategy.

Before introducing the m_{klt} variables and U, M constraints we solved Smith's first encoding (constraints S,V,Y) with local search. With 32136 vari-

---

[5] We thank Mats Carlsson for confirming this observation with a SICStus FD implementation.

ables and 90844 constraints (after fixing the hosts), this encoding was much larger. Nevertheless, WSAT(OIP) solved it in a few minutes.

Related Work. Recently, Hooker and Osorio [76] introduced a framework called Mixed Logical/Linear Programming (MLLP). They apply MLLP to the progressive party problem and compare the experimental results to a MIP encoding solved with CPLEX. Instead of using the two-stage factoring presented above, Hooker and Osorio encode both stages of the problem using a compact representation with the number of host boats as minimization objective. Their approach (MLLP) can solve problems of up to 10 boats and 4 time periods to optimality within several hours (for comparison, the original problem has 29 boats and 6 time periods); MILP is reported [76] to find optimal solutions up to 8 boats and 4 periods without manual intervention.

5.2 The ACC Basketball Scheduling Problem

In the second case study in timetabling/scheduling, we investigate a difficult problem from sports scheduling that was recently studied and solved by Nemhauser and Trick [115], the scheduling of the Atlantic Coast Competition in basketball (*ACC Basketball 97/98*). The previous approach by Nemhauser and Trick (N&T) involved a domain-specific problem factorization together with a mix of integer programming and explicit enumeration leading to a solution that was accepted by the ACC.

Here, we investigate an integer local search approach to the ACC problem that works directly from a monolithic 0-1 integer linear program and includes all of the documented constraints [115, 141] of the original problem.

With respect to the experimental results, integer programming and explicit enumeration have been reported [115] to find a set of schedules within around 24 hours on a modern workstation (the approach is exact and finds all solutions to the problem modulo certain restricting assumptions, i.e. the particular mirroring scheme).[6] More recently, a very efficient approach to the problem has been reported by Henz [69] that applies constraint programming to a problem factorization by Schreuder [130] (similar to the factorization used in [115]).

In contrast to the previous approaches, the integer local search approach uses no problem factorization, but still finds solutions that are competitive with the official timetable [115] with respect to several optimization criteria. In summary, the results of the case study are:

(i) Both modeling and solving of the ACC problem can be accomplished using a monolithic IP representation. Requiring solutions to be at least

---

[6] Note that much shorter times have been reported for finding a first solution (Michael Trick, personal communication). In the last stage, 300 million schedules are generated and filtered.

as good as the official timetable with respect to all optimization criteria given in [115], solutions are found in 30 minutes (on average) by $\text{W}_\text{SAT}(\text{OIP})$. This is an exciting result for local search as the problem (with a potential search space of 2^{1773} variable assignments) has only 87 solutions!

(ii) A general purpose heuristic, $\text{W}_\text{SAT}(\text{OIP})$, can solve a real instance of a dense double-round-robin (DDRR) scheduling problem.

(iii) In a double round robin competition, the second half of the schedule typically mirrors the first. However, to comply with given team pairing requirements, such perfect mirroring is not always possible: To handle additional pairing requirements, the previous approaches based on factorization resort to swapping slots of the schedule, which is not possible when conflicting pairing requirements exist. To deal with conflicting pairings, we present *minimal distortion mirroring*, a new approach in which only few pairings are swapped while the basic mirroring scheme is preserved.

5.2.1 Double Round Robin Scheduling

In a *Double Round Robin (DRR)* sport competition, which is a popular scheme in many sports, every team t plays against every other team exactly twice during the competition, once at home (the place of t) and once away.

There are two types of sports schedules: temporally dense and temporally relaxed. In temporally dense double round robin scheduling (DDRR) like the ACC competition, the number of slots (time periods in which games may take place) is almost equal to the number of games that each team must play. If there is an even number n of teams, a DDRR schedule has $2(n-1)$ slots. If n is odd, there are $2n$ slots in which $n-1$ teams play and one team is bye.

Contrary to temporally relaxed schedules where local improvement heuristics appear to be used frequently (e. g. [45]), a brief survey in [115] attests for temporally-dense schedules that "while some local improvement heuristics have been found, they tend to be rather limited in scope and heavily dependent on finding good initial solutions." Hence, by using a (i) domain-independent (ii) heuristic which starts from (iii) a random initial solution, our approach takes several steps in one.

As with most other real sport scheduling problems, the ACC 1997/98 problem is constrained by a wide variety of requirements and objectives. For example, it is desired to have a large separation between the two games of a pair of teams. If the criterion is to minimize the maximal distance between any two teams, the minimal separation is half the number of teams. In this situation, the problem is typically simplified by requiring that the pairings in the first and the second half be identical, except that the places of the games are reversed. If the timetable meets this condition, it is said to be *perfectly mirrored.* In the ACC 97/98 problem, individual team pairing constraints prohibit a *perfect* mirroring.

Another important aspect concerns the satisfaction of the local spectators who prefer not too few and not too many local games in a sequence. Therefore, the succession of home and away matches needs to be altered frequently. The requirements of the ACC given in [115] with respect to the allowed sequences of home/away and bye are intricate and rule out a direct application of previous work on double round robin tournaments [25, 130]. Further, there are constraints that no team should face the particularly strong teams in immediate succession.

The third aspect concerns broadcasting, as television networks require a stream of "high quality" games and have additional requirements when the most popular pairings should occur. Since teams return home after almost every away game in the ACC, there are no travel constraints.

Because the problem characteristics change if details of the specification are omitted, it is unavoidable to present the entire list of constraints (as will be shown, simplified versions of the problem are in fact very easy for integer local search). The following section therefore presents the complete "laundry-list" [115] of constraints.

5.2.2 Problem Specification of ACC97/98

The ACC in basketball consists of nine universities: Clemson (Clem), Duke, Florida State (FSU), GeorgiaTech (GT), Maryland (UMD), North Carolina (UNC), North Carolina State (NCSt), Virginia (UVA), and Wake Forst (Wake). The problem is to find a 18 slot DDRR for the period of nine weeks (12/31/97, a Wednesday to 3/1/89, a Sunday), such that in each week there is a weekday and a weekend game. In the following description of the requirements, we follow exactly the presentation of Henz [69] and distinguish 'requirements' from 'optimization criteria'. The requirements can be viewed as a minimal set of constraints to satisfy, whereas the optimization criteria may be met in different ways since they are generally in conflict. The requirements are the following:

R0. *Double round robin.* The teams play a temporally dense double-round robin competition.

R1. *Return match separation.* The teams wish return their games as separate as possible (i.e. if a at b in slot i, then b at a at $i + D$ for suitably large D). The measure is to maximizing the minimum distance. The requested separation between games between two teams also holds for byes: No team wants its two byes too close together. The minimal temporal distance between *first leg* and corresponding *return match* must be 7 slots. Considering that UNC plays Duke in slot 11 an 18 (see requirement 9 below) 7 is the maximal value for this minimal distance.

R2. *No two final aways.* No team can play away in both last slots.

R3. *Home/Away/Bye pattern constraints.* No team may have more than two away matches in a row. No team may have more than two home matches in a row. No team may have more than three away matches or byes in a row. No team may have more than four home matches or byes in a row. Similar conditions hold for consecutive weekend slots. No team may have more than two away matches on subsequent weekends. No team may have more than two home matches on subsequent weekends. No team may have more than three away matches or byes on subsequent weekends. No team may have more than three home matches or byes on subsequent weekends.

R4. *Weekend pattern.* Of the weekends, each team plays four at home, four on the road, and one bye.

R5. *First weekends.* Each team must have home matches or byes at least on two of the first five weekends.

R6. *Rival matches.* Every team except FSU has a traditional rival. The rival-pairs are Duke-UNC, Clem-GT, NCSt-Wake, and UMD-UVA. In the last slot, every team except FSU plays against its rival, unless it plays against FSU or has a bye.

R7. *Popular matches in February.* The following pairings must occur at least once in slots 11 to 18: Wake-UNC, Wake-Duke, GT-UNC, and GT-Duke.

R8. *Opponent ordering constraints.* No team plays in two consecutive slots away against UNC and Duke. No team plays in three consecutive slots against UNC, Duke and Wake (independent of home/away).

R9. *Other idiosyncratic constraints.* UNC plays its rival Duke in the last slot and in slot 11. UNC plays Clem in the second slot. Duke has a bye in slot 16. Wake does not play home in slot 17. Wake has a bye the first slot. Clem, Duke, UMD and Wake do not play away in the last slot. Clem, FSU, GT and Wake do not play away in the first slot. Neither FSU nor NCSt have a bye in the last slot. UNC does not have a bye in the first slot.

R10. A small set of additions to the original description have been published very recently [141]: Every team must have an H in the first three slots. Every team must have an H in the last three slots. Wake is bye in the first slot and must end AH.

Optimization criteria. There are several additional criteria the ACC requires of a time-table [115]. As Henz notes [69], some of these optimization criteria are conflicting, so the best one can hope for are Pareto-optimal solutions. The goal of the integer local search approach is to find solutions that are at least as good as the official 97/98 schedule.

O1. Avoid two opening aways. The number of teams that play away in the first two slots should be small. We denote this number by *OAA*.

O2. Good slots in February. Table 5.3 classifies (a) weekday and (b) weekend games into A-games, B-games and bad games (represented by 0). If a slot

Table 5.3. Game quality

home		away								
		Clem	Duke	FSU	GT	UMD	UNC	NCSt	UVA	Wake
	Clem	0	0	0	0	0	B	0	0	0
	Duke	0	0	0	B	A	0	0	B	B
	FSU	0	0	0	0	0	0	0	0	0
	GT	0	B	0	0	B	A	0	0	B
	UMD	0	A	0	B	0	A	0	B	0
	UNC	B	A	0	B	B	0	0	0	0
	NCSt	0	B	0	0	0	B	0	0	B
	UVA	0	B	0	0	0	0	0	0	B
	Wake	0	B	0	B	0	B	B	0	0

home		away								
		Clem	Duke	FSU	GT	UMD	UNC	NCSt	UVA	Wake
	Clem	0	0	0	0	0	A	0	0	0
	Duke	0	0	0	B	A	A	0	B	B
	FSU	0	0	0	0	0	0	0	0	0
	GT	0	0	0	0	B	0	0	0	B
	UMD	0	0	0	0	0	0	0	0	0
	UNC	0	0	0	B	B	0	0	0	0
	NCSt	0	B	0	0	0	B	0	0	B
	UVA	0	B	0	0	0	0	0	0	0
	Wake	0	B	0	B	0	B	B	0	0

Weekday Games   Weekend Games

contains at least one A-game or at least two B-games, it is called an A-slot. If a slot is not an A-slot and contains at least one B-game, it is a B-slot. All other slots are bad slots. In February (slots 11 through 18), the A-slots should be maximized and the bad slots minimized.

O3. Home/Away/Bye pattern criteria. The number of occurrences of three subsequent home matches or byes should be small (HB_3). Similarly, the number of occurrences of three subsequent away matches or byes (AB_3). And again, for weekends the same criteria should hold (HB'_3, AB'_3).

Any schedule can be rated according to the above optimization criteria, summarized by a vector of 7 numbers. The official timetable computed by Nemhauser and Trick meets these optimization criteria as follows:

OAA	HB_3	AB_3	HB'_3	AB'_3	bad	A-slots
1	4	3	5	4	2	3

5.2.3 Integer Local Search Formulation

In this section, a 0-1 integer linear local search model will be developed to state all the requirements. To simplify the model, we number the teams in the order given above. For the integer local search model, we introduce indices i and j that range over teams $\{1 \ldots 9\}$ and t which ranges over slots $\{1 \ldots 18\}$. The binary decision variables are x_{ijt} and $x_{ijt} = 1$ iff team i plays as guest of team j in slot t, for all $1 \leq i, j \leq 9$. Additionally, a team index of 0 is used to express byes, e.g. $x_{i0t} = 1$ iff team i is bye in slot t. Similarly, x_{iit} encodes homes, i.e. $x_{iit} = 1$ iff team i plays at home in slot t.

We present the constraints from the most general ones to the more specific and finish with the idiosyncrasies of the ACC97/98 season.

R0. The following constraints implement the basic double round robin scheme, requirement R0.

Every team plays at exactly one place (or is bye) in every slot.

$$\sum_{0 \le j \le 9} x_{ijt} = 1, \quad \text{for all } i > 0, t \tag{5.1}$$

Every team is visited by at most one team in every slot.

$$\sum_{0 < i \ne j} x_{ijt} \le 1, \quad \text{for all } j > 0, t \tag{5.2}$$

All pairings are consistent.

$$x_{ijt} \rightarrow x_{jjt}, \quad \text{for all } 0 < i \ne j \text{ and } t \tag{5.3}$$

Double round robin: Every team plays every other team once away (and once at home which is implied in combination with (5.3)).

$$\sum_{t} x_{ijt} = 1, \quad \text{for all } i, 0 < j \ne i \tag{5.4}$$

R3. The following constraints restrict the allowed game sequences (Home/ Away/Bye pattern constraints), and need to be duplicated for the weekend slots.

Treating bye as away, no more than 2 away games in a row.

$$\sum_{s=t...t+2} \overline{x_{iis}} \le 2, \text{ for all i, } 1 \le t \le T-2 \tag{5.5}$$

Treating bye as home, no more than 3 home games in a row.

$$\sum_{s=t...t+3} (x_{iis} + x_{i0s}) \le 3, \quad \text{for all i, } 1 \le t \le T-3 \tag{5.6}$$

Treating bye as away, no more than 2 home games in a row.

$$\sum_{s=t...t+2} x_{iis} \le 2, \quad \text{for all i, } 1 \le t \le T-2 \tag{5.7}$$

Formulating the Mirroring Scheme.

R1. To ease comparison with the previous approaches [115, 69], we will use the same mirroring scheme throughout our experiments. Due to requirement R9 which states that UNC and Duke meet in slots 11 and 18, perfect mirroring is not possible. For this reason, the N&T approach resorts to a mirroring scheme that switches slots 9 and 11 and obtain a mirroring scheme in which the minimal distance between any pair of teams is 7. The N&T approach also switches slots 8 and 9 (although no obvious constraint enforces this), and arrives at a set of mirrored slots of

$$M_{NT} = \{(1,8),(2,9),(3,12),(4,13),(5,14),(6,15),(7,16),(10,17),(11,18)\}.$$

Note that fixing a conflict in a perfect mirroring scheme by switching entire slots also affects all the other pairings; all teams meeting in slot 1 will meet again in slot 9—only because of the UNC–Duke meeting. We will discuss an alternative to this scheme in Section 5.2.7. The constraints that enforce the mirroring can be formulated as:

Mirror return games.

$$x_{ijs} \rightarrow x_{jit} \quad \text{for all } i \neq j \text{ and } (s,t) \in M_{NT} \qquad (5.8)$$

Mirror byes.

$$x_{i0t} \rightarrow x_{i0s} \quad \text{for all } i \neq j \text{ and } (s,t) \in M_{NT} \qquad (5.9)$$

ACC Specific Requirements.

R2, R4, R5. The following restrictions are more specific to the ACC and concern unliked sequences of H/A/B.

No team finishes AA.

$$x_{i,i,T-1} + x_{i,0,T-1} + x_{i,i,T} + x_{i,0,T} \geq 1, \text{ for all } i \qquad (5.10)$$

Of the 9 Saturdays, each team plays four at home, four on the road, and one bye.

$$\sum_{t=2,4,\dots,T} x_{iit} = 4, \qquad \sum_{t=2,4,\dots,T} x_{i0t} = 1 \qquad (5.11)$$

Each team must be home or bye at least on two of the first five weekends.

$$\sum_{t=1\dots5} x_{iit} + x_{i0t} \geq 2, \quad \text{for all } i \qquad (5.12)$$

R6, R7, R8, R9, R10. Finally, the team-specific requirements of the ACC97/98. We do not present the constraints here as for most of these requirements the the constraint encoding is straightforward. An exception is constraint R6 (rival games), for which an equivalent formulation can be used, i.e., three of the four rival games must be played in the last slot.

Optimization Criteria. The optimization criteria are modeled here as hard constraints to ensure that all resulting solutions will be at least as good as the official schedule. Soft constraints could be used alternatively, of course.

O1, O2, O3. Modeling O1 is straightforward:

Avoid opening away/away (no more than 1 team).

$$\sum_i (x_{i,i,1} + x_{i,0,1} + x_{i,i,2} + x_{i,0,2}) \geq 8 \tag{5.13}$$

To formulate O2, additional variables need to be introduced for all slots in February; $q_{tr} = 1$ if the quality of slot $t \in Feb$ is r, where $r \in \{0,1,2\}$, $r = 0$ represents a bad slot, $r = 1$ a B, and $r = 2$ an A slot, respectively (of course, $\sum_r q_{tr} = 1$ for all t). An example from the constraints that link the x and q variables:

Given the game quality matrix G^w for weekdays, every weekday slot in February that is marked as an A slot must contain at least one A or two B games.

$$2 * q_{t2} \leq \sum_{i \neq j} G_{ij}^w * x_{jit}, \quad \text{for all weekdays } t \in Feb \tag{5.14}$$

Similarly for B slots and for week ends.
Ensure the N/T quality level for game qualities.

$$\sum_{t \in Feb} q_{t2} \geq 3 \quad \sum_{t \in Feb} q_{t0} \leq 2 \tag{5.15}$$

Similarly, to ensure the Home/Away/Bye sequences stated in O3, additional variables are introduced for every team and time point t_0 that state when long unliked sequences occur starting at t_0. For the full list of constraints, the reader is referred to Appendix A.

5.2.4 Redundant Constraints

There are certain simple truths about DDRR schedules that are implicit in any DDRR encoding and that can be explicated in order to improve the operational performance of integer local search.

Such *redundant constraints* are often employed in other domain-independent frameworks to improve a representation with respect to the operational behavior of a solver when applied to it. In constraint programming, for example, redundant constraints can strengthen the propagation. In integer linear programming, redundant constraints are used to tighten the LP relaxation. Also, redundant constraints have recently been used for local search by Kautz and Selman [94] in SAT planning models. The following redundant constraints are used in the DDRR model.[7]

Every team is bye twice.

$$\sum_t x_{i0t} = 2, \quad \text{for all } i \tag{5.16}$$

One team is bye in every time slot (holds for an odd number of teams only).

$$\sum_i x_{i0t} = 1, \quad \text{for all } t \tag{5.17}$$

Half the teams are home in each slot.

$$\sum_i x_{iit} = 4, \quad \text{for all } t \tag{5.18}$$

5.2.5 Previous (Multi-stage) Approaches

The previous strategies to solve the ACC problem [115, 69] factor the problem into several stages, following earlier approaches to sports scheduling by Cain [25], Schreuder [130], and others. Each stage of the problem is solved individually and ensures that a subset of the full set of constraints are met. Solving all stages in sequence yields complete timetables that meet all the constraints.

(i)   The first stage of the multi-stage approach generates so-called *patterns*: A feasible pattern is a sequence of H/A/B (one letter for each slot of the DDRR) which meets the particular mirroring scheme and all given H/A/B constraints.

(ii) The second stage produces so-called *pattern sets*: a number of patterns (as many as there are teams) are selected from the collection of patterns into a pattern set. The chosen patterns of each pattern set must meet the condition that for every slot, there must be four patterns with an H, four with an A and one B. Additionally, one can require a pattern set to minimize the number of less preferred patterns [115] (e. g. requirement O1).

(iii) The third stage, finally, renders the complete *timetables*. A timetable is computed on the basis of a given pattern set by assigning one team to each pattern in the set. A timetable is feasible if all problem constraints are met. In [115] ACC timetables

---

[7] We notice that there is an interesting connection to the pattern set approaches: The redundant constraints (5.17) and (5.18) express exactly the constraints on grouping patterns into pattern sets.

are computed with an additional intermediate stage that assigns team placeholders to pattern sets first.

To solve the problem, the N&T approach [115] uses explicit enumeration in stage (i), integer programming in stage (ii), and integer programming and explicit enumeration in stage (iii). This indicates the difficulty of solving the problem.

Henz [69] uses constraint programming to solve the individual stages. The CP approach, implemented in Oz [137], turns out to be significantly more efficient than integer programming with explicit enumeration, primarily because the time-consuming final explicit enumeration phase is more efficiently accomplished by using constraint propagation.[8]

5.2.6 Experimental Results under Varied Constrainedness

In this section, we describe the experimental results of integer local search. All above constraints were modeled with the AMPL algebraic modeling language. After AMPL preprocessing, the constraints were handed to WSAT(OIP) in expanded form using an AMPL-WSAT(OIP) interface (Appendix A contains the full AMPL model).

Table 5.4 shows a timetable found by WSAT(OIP) for the entire set of constraints given above (R0–R10,O1–O3). The quality of the timetable improves the official timetable given in [115] with respect to several formalized quality measures (note that additional *informal* considerations actually lead to the selection of the official timetable [115]).

OAA	HB_3	AB_3	HB'_3	AB'_3	bad	A-slots
1	3	1	5	4	0	4

The initial experiments of WSAT(OIP) were carried out before all requirements were available and lead to promising results.[9] As more and more constraints unfolded, the solution times increased. To capture this behavior, we next present experimental results for a sequence of problems with increasing constrainedness.

Table 5.5 investigates the scaling of WSAT(OIP) with increasing constrainedness. The problem instances start from general constraints of the DDRR scheme and successively incorporate more specific constraints. The number of solutions to the problem was computed using the Oz constraint program by Henz [69].[10] We additionally ran an IP branch-and-bound procedure (CPLEX 5.0) on the given problems to obtain an estimate of its capabilities for the monolithic 0-1 ILP.

---

[8] Oz is publicly available from http://www.ps.uni-sb.de/oz/.

[9] I thank Michael Trick and George Nemhauser for sharing the requirements at an early stage.

[10] I thank Martin Henz for sharing the Oz program. Note that our tightness level t5 corresponds to tightness 0 reported in [69].

Table 5.4. A Schreuder-timetable computed by WSAT(OIP) from an AMPL model of the constraints R1–R10, O1–O3. The format is the same as the one used in [130]. +1 means home against 1, −1 means away at team 1, 0 means 'bye'. The particular run took 150s.

		slots																	
		1	2	3	4	5	6	7	8	9	10	11	12	13	14	15	16	17	18
	Clem 1	+8	−6	+4	−3	−2	+5	−9	−8	+6	+7	0	−4	+3	+2	−5	+9	−7	0
	Duke 2	−5	+7	+9	−4	+1	−3	0	+5	−7	+8	−6	−9	+4	−1	+3	0	−8	+6
	FSU 3	+6	−9	−7	+1	0	+2	−8	−6	+9	−5	+4	+7	−1	0	−2	+8	+5	−4
	GT 4	+7	−8	−1	+2	−5	0	+6	−7	+8	−9	−3	+1	−2	+5	0	−6	+9	+3
	UMD 5	+2	0	−6	+9	+4	−1	+7	−2	0	+3	−8	+6	−9	−4	+1	−7	−3	+8
	UNC 6	−3	+1	+5	−8	+7	+9	−4	+3	−1	0	+2	−5	+8	−7	−9	+4	0	−2
teams	NCSt 7	−4	−2	+3	0	−6	+8	−5	+4	+2	−1	+9	−3	0	+6	−8	+5	+1	−9
	UVA 8	−1	+4	0	+6	−9	−7	+3	+1	−4	−2	+5	0	−6	+9	+7	−3	+2	−5
	Wake 9	0	+3	−2	−5	+8	−6	+1	0	−3	+4	−7	+2	+5	−8	+6	−1	−4	+7

Table 5.5. Experimental results for increasing constrainedness. Columns report the constraints added, the 'tightness' level, the number of variables and constraints of the problem, the total number of solutions ('unkn' means unknown). The WSAT(OIP) columns report on runtimes with (+red.) and without (−red.) redundant constraints (5.16)-(5.18). All runtimes are time to first solution on an Intel Pentium Pro 300Mhz ('−' means no experiment performed, CNS=could not be solved in 12h runtime). For WSAT(OIP), runtimes are averaged over 50 runs (+red) 20 runs (−red), and 10 runs†. The maximal standard error in column +red is 13%.

constraints added	cmnt	lev	n	m	#sols	WSAT(OIP) −red.	WSAT(OIP) +red..	CPLX 5.0
(5.1)-(5.4)	DDRR	t0	1620	1737	unkn	1s	0.1s	183s
(5.5)- (5.7)	H/A/B	t1	1620	2286	unkn	15s	0.4s	649s
Weekends	H/A/B	t2	1620	2520	unkn	79s	1s	873s
(5.8)- (5.9)	Mirror	t3	1620	3249	>1e5	27s	9s	174m
(5.10)-(5.12)	R2,4,5	t4	1620	3285	−	631s	162s	CNS
Idiosync.	R6-R9	t5	1339	3052	321	245m†	1664s	CNS
(5.13)	O1	t6	1335	3047	321	−	1484s	CNS
(5.14)-(5.15)	O2	t7	1359	3069	272	−	2128s	CNS
Opt. H/A/B	O3	t8	1773	3466	88	−	2847s	CNS
Recent add.	R10	t9	1773	3479	87	−	1798s	CNS

Discussion of Results. The approach by Nemhauser and Trick [115] yielded a complete set of solutions for the problem of tightness t9 with a turn-around time of 24 hours. The multi-stage constraint program [69] is currently the most efficient approach and finds all solutions to the full problem, requiring only a few minutes (depending on the exact tightness of the problem). Interestingly, opposed to local search, runtimes of the CP approach to find the first solution tend to decrease the tighter the problem is constrained. Conversely, it occasionally exhibits problems on loosely constrained variants of the problem, when the deterministic search happens to enter large subtrees that do not contain solutions. For example, tightness t4 does not render a solution within hours. When the number of generated patterns (first stage) is large, the CP approach can also run into difficulties if useless pattern sets are produced first.

Integer Local Search. The general observation from Table 5.5 is that as the the number of solutions decreases, runtimes of WSAT(OIP) increase, as is to be expected with local search. What is surprising, however, is that solutions are still found when the constrainedness has reached a level that rules out all but 87 solutions! Another observation is that the redundant constraints play an important role for local search in this problem, in particular as the constrainedness is increased (thus we did not conduct experiments above tightness level t5 without redundant constraints). More evidence is provided by the fact that posting additional constraints which do not change the number of solutions tends to decrease runtime.

Throughout the experiments, WSAT(OIP) was run in single-solution mode, although a multi-solution search would arguably be more appropriate for a comparison to the previous approaches. A multi-solution search would involve continuing search after having found a feasible solution, while avoiding previously generated timetables. Possibly, the search could be encouraged to remain 'close' to the previously found solutions.

Comparison. Since local search is incomplete, it is not possible to search for the complete set of solutions and then stop. On the other hand, to be able to solve the problem, restrictive assumptions are made in the N&T and CP formulations (the mirroring scheme) which rule out a number of solutions to the problem that would otherwise be considered perfectly acceptable. The main advantage of the completeness of N&T and CP for this problem is thus their ability to detect when no solutions exist to the constraints.

The main disadvantage of a *pure* local search approach is thus that there is no response if no feasible solution exists to a particular model. However, since the problem is formulated using integer constraints, LP relaxations can be employed: When tightening the constraints leads to IP infeasibility, the corresponding LP will sometimes also be infeasible. LP infeasibility can normally be determined efficiently by linear programming. For example consider tightening constraint (5.5) to "treating bye as away, no more than 1 away game in a row"; the LP relaxation of the resulting model is proved infeasible by CPLEX in 75 seconds. Also, existing LP presolving can sometimes prove infeasibility. For instance, we accidentally swapped incorrect slots of the mirroring scheme and obtained an infeasibility warning from the AMPL presolver instantly.

Parameters. WSAT(OIP) was run with parameters $t = 2$, $p_{noise} = 0.01$, Maxmoves=2e6, $p_{zero} = 0.9$. Note that the runtime variation due to parameter variations turned out to be small (similar to the standard error for 20 runs). For example, changing parameters to $p_{noise} = 0.2$, or additionally turning off history yielded similar results for these problems. However, a rigorous experimental analysis is beyond the scope of this monograph.

To obtain an estimate of the capabilities of IP branch-and-bound for the monolithic 0-1 ILP, we also ran CPLEX on the given problems. All results report on standard parameter settings. CPLEX was further tried with a feature ('sosscan') that identifies special ordered sets (sos type 3, i.e. a set of binary variables that appear in a less-than or equality constraint with +1 coefficients and an RHS value of +1), to apply special branching strategies. Despite 20% of the constraints being of this type, performance degraded with this option. We also attempted other option changes which did not improve performance, e.g. changing the branching direction (by setting the branch variable first to one in order to improve propagation), and "strong branching", recommended for hard pure integer programming problems. Note that we cannot rule out that a different IP model or other parameter settings might improve performance.

5.2.7 Minimal Distortion Mirroring

As mentioned earlier, in order to comply with the given team pairing constraints that are in conflict with the mirroring scheme, the previous approaches swap entire slots. Swapping slots, however, is problematic when several preassigned team pairings are in a conflict: For example, suppose that the team pairing $\{1, 2\}$ is required in slots t_1, t_2 while pairing $\{3, 4\}$ is required in slots t_1, t_3 (neither t_2 nor t_3 mirroring t_1). In this case, if $t_2 \neq t_3$, swapping slots is not possible because two swaps would be required that are inconsistent with each other. To handle this case, we will resort to swap individual opponents instead of entire columns.

We will refer to the slot mirroring some slot t in the original mirroring scheme as t'. Consider the case in which the input constraints fix the team pairing $\{a, b\}$ for slots $t_1 < t_2$ wherein $t_1' \neq t_2$ (i.e. t_2 does not mirror t_1). Now instead of swapping column t_1' and t_2 like before, we limit the distortion of the original mirroring scheme. The idea is to pick two suitable teams c_1 and c_2 (called *sweeper* teams) and make the following changes: Relax all mirror constraints involving a, b, c_1, c_2 for the slots t_1, t_1', t_2, t_2' and add constraints to fix the pairings of the sweeper teams. Figure 5.1 illustrates the situation for the case of t_1 and t_2 both being in the first half of the season.

In this *minimal distortion mirroring*[11], the basic mirror pattern is mostly preserved and additional team pairing requirements can be included. The distance to return games remains the distance of the perfect mirroring for all pairings except $(i, j) \in \{a, b, c_1, c_2\}^2, i \neq j$, thereby improving the mean distance between a game and its return game.

Modeling. In order to formulate minimal distortion mirroring, we employ additional variables to select the teams c_1 and c_2. For each team $1 \leq i \leq 9$, the binary variable $y_i = 1$ iff i is a sweeper team (of course, teams a and b

---

[11] I thank Martin Henz for coining this term in a discussion.

t_1	$\cdots$	t_2.		t_1'	$\cdots$	t_2'
$\{a,b\}$		$\{a,b\}$	$\Longrightarrow$	$\{a,c_1\}$		$\{a,c_2\}$
$\{c_1,c_2\}$		$\{c_1,c_2\}$		$\{b,c_2\}$		$\{b,c_1\}$

Figure 5.1. Minimal distortion mirroring.

cannot be sweeper teams). First, to achieve a *minimal* distortion, we require the number of sweeper teams to be minimal, i.e. $\sum_i y_i = 2$.

Next, we need to reformulate the mirror constraints, choosing a perfect mirror $M_p = \{(s, s+9) : 1 \leq s \leq 9\}$. While we mirror byes exactly as before (5.9), the following constraints substitute (5.8).

Mirror return games as usual except in slots t_1, t_1', t_2, t_2'.

$$x_{ijs} \to x_{jit} \quad \text{for all } i \neq j \text{ and } (s,t) \in M_p; \ s,t \notin \{t_1, t_1', t_2, t_2'\} \quad (5.19)$$

In slots t_1, t_1', t_2, t_2', mirror return games as usual except for games against conflict teams (a and b) and the (current) sweeper teams.

$$y_i \vee y_j \vee (x_{ijs} \to x_{jit}) \quad \text{for all } i \neq j; \ i,j \neq a,b$$
$$\text{and } (s,t) \in M_p; \ \{s,t\} \cap \{t_1, t_1', t_2, t_2'\} \neq \emptyset \quad (5.20)$$

In slot t_1 and t_2, if i and j are sweeper teams, they must meet.

$$(y_i \wedge y_j) \to (x_{ijt} \vee x_{jit}) \quad \text{for all } i < j; \ i,j \neq a,b; \ t \in \{t_1, t_2\} \quad (5.21)$$

In slot t_1' and t_2', if i is a sweeper team, it must meet a conflict team.

$$y_i \to \bigvee_{j \in \{a,b\}} (x_{ijt} \vee x_{jit}) \quad \text{for all } i \neq a,b; \ t \in \{t_1', t_2'\} \quad (5.22)$$

Note that (5.21) and (5.22) are still linear inequalities. Given the above model of the ACC problem (R1–R9), we obtain the timetable in Table 5.2.7 together with the automatically assigned sweeper teams from WSAT(OIP) (average runtime was not measured in this experiment).

In contrast to the M_{NT} mirror, which has an average distance between pairings of 8.11, the minimal distortion schedule achieves a distance of 8.80.[12] The quality vector of timetable 5.2.7 is:

OAA	HB_3	AB_3	HB_3'	AB_3'	bad	A-slots
0	4	3	5	4	1	3

---

[12] Of the $9 \times 9 = 81$ return games (counting byes), only 8 have a distance of 7 from their first leg, all others have a distance of 9.

Table 5.6. A minimal distortion timetable from WSAT(OIP), meeting the requirements R1–R9. Preassigned games that conflict with the regular mirror are shown in boldface. To correct the conflict, two sweeper teams are selected, which happens automatically when the described model is solved. Games that do not follow the standard mirroring scheme are drawn in boxes (as well as the sweeper teams).

		slots																	
		1	2	3	4	5	6	7	8	9	10	11	12	13	14	15	16	17	18
			t_1'							t_2'	t_1								t_2
Clem	1	+2	[−6]	−8	0	+4	−7	−9	+5	[+6]	−2	[−3]	+8	0	−4	+7	+9	−5	[+3]
Duke	2	−1	[+3]	+5	−9	+7	−4	0	+8	[−3]	+1	[−6]	−5	+9	−7	+4	0	−8	[+6]
FSU	3	+8	[−2]	+7	+4	−9	+6	−5	0	[+2]	−8	[+1]	−7	−4	+9	−6	+5	0	[−1]
GT	4	+5	−8	+9	−3	−1	+2	−6	+7	0	−5	+8	−9	+3	+1	−2	+6	−7	0
UMD	5	−4	+9	−2	−7	+6	0	+3	−1	−8	+4	−9	+2	+7	−6	0	−3	+1	+8
UNC	6	−7	[+1]	0	+8	−5	−3	+4	−9	[−1]	+7	[+2]	0	−8	+5	+3	−4	+9	[−2]
NCSt	7	+6	0	−3	+5	−2	+1	+8	−4	+9	−6	0	+3	−5	+2	−1	−8	+4	−9
UVA	8	−3	+4	+1	−6	0	+9	−7	−2	+5	+3	−4	−1	+6	0	−9	+7	+2	−5
Wake	9	0	−5	−4	+2	+3	−8	+1	+6	−7	0	+5	+4	−2	−3	+8	−1	−6	+7

5.3 Conclusions

This chapter has studied two hard timetabling/scheduling problems, The Progressive Party Problem and scheduling of the ACC97/98 basketball conference. Both problems can be formulated as 0-1 integer linear programs. It has reported the first ILP model for the ACC problem that we are aware of and improved the existing ILP model for progressive party.

To the best of our knowledge, no previous techniques have been reported to solve either of the problems from a given 0-1 ILP representation (Smith *et al.* [136] report a number of unsuccessful attempts using integer programming branch-and-bound). From the viewpoint of integer programming, the contribution of this chapter is thus to demonstrate that both problems can be solved in a 0-1 integer constraint encoding using a general solver (WSAT(OIP)).

From the applications viewpoint, we have shown in the first case study that the progressive party problem can be solved efficiently and robustly using WSAT(OIP), i. e. the strategy scales gracefully with increasing constrainedness of the instances (the original study [136] investigated only one instance).

The second case study has demonstrated that integer local search is able to find solutions to ACC97/98, a difficult and complex timetabling problem. We have shown that our approach yields solutions competitive with the official timetable (reported in [115]). We also presented minimal distortion mirroring, a mirroring scheme that can still handle the situation when team pairing requirements conflict with swapping time slots of a mirroring scheme, the strategy employed in previous (factorization) approaches.

Additionally, the second case study in this chapter has studied the behavior of WSAT(OIP) across a range of increasingly tight problems. While we have seen that the runtime of local search does increase on very tight problems, we have demonstrated that even extremely tight problems can be solved by WSAT(OIP). Moreover, we have shown that adding redundant constraints can help local search to find solutions more quickly.

6. Covering and Assignment

This chapter investigates two integer optimization problems, radar surveillance and course assignment. For both problems, the 0-1 OIP encoding is straightforward. Structurally, the problems are extensions of set covering and generalized assignment, respectively. The first problem stems from an industrial project at the Swedish Institute for Computer Science (SICS), while the second problem arose from an operating application at the Universität des Saarlandes. Both studies in this chapter will focus on performance variation of integer local search and IP branch-and-bound with increasing problem size.

Further, using the radar surveillance problems, we will perform experiments to determine the impact of the OIP representation on performance. The experiments demonstrate that the WSAT(OIP) method critically depends on the soft constraint representation using constraint bounds.

6.1 Radar Surveillance Covering

The problem considered in this section is related to the classic NP-hard set-covering problem (see (SCP) in the Introduction). It extends set covering by complicating side-constraints that are specific to the radar domain and which prevent a direct application of domain-specific heuristics from the literature. On the other hand, its particular structure is well-suited for an encoding into OIP and makes it an intersting test case for integer local search.

6.1.1 Problem Description and Formulation

The case study and its basic modeling originate from a project currently carried out at the Swedish Institute of Computer Science (SICS) [24]. The goal is to plan radar surveillance of a geographic area. As customary in the radar surveillance domain, the area is divided into *hexagonal cells*. As part of the problem statement, a number of radar stations are given that are located in fixed cells on the map. The problem is to find a static plan that determines for every cell c by which radar stations c is observed, subject to the constraints that each cell be observed by at least three stations. Figure 6.1 gives an illustration.

Each radar station can divide its signal scope circle into six sectors and can vary the signal strength in each sector independently from zero to some given maximum distance d_{max}. Aside from some insignificant cells, the majority of the cells must be covered by 3 radar stations (desired coverage 3) and all coverage beyond this is to be minimized (*over-coverage*). Small over-coverage is desired for economic reasons as well as for reasons of detectability, i. e. radar can more easily be detected in areas with a high exposure.

Because of the placement of stations, some cells cannot (physically) be covered by at least three stations and hence must be covered by as many stations as possible (and can then be factored out from the problem). In the original model, a radar station can be switched on to cover only the cell that it is located in. It always covers it provided it is switched on for some sector.

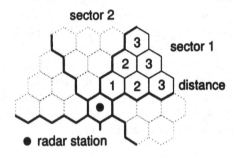

Figure 6.1. Radar map with hexagonal cells.

The problem can be modeled by the following over-constrained 0-1 integer program. For every combination of radar unit u, sector $1 \leq s \leq 6$ and possible observation distance $1 \leq d \leq d_{max}$, a Boolean variable σ_{usd} is introduced. Variable $\sigma_{usd} = 1$ if and only if station u is switched on in sector s at distance d. The set of all cells that station u reaches in sector s at distance d is denoted by C_{usd}. The over-constrained integer program (OIP) model is as follows.

Cover each cell. There are significant and insignificant cells. While insignificant cells should not be observed, significant cells must be covered by at least three stations.

$$\sum_{c \in C_{usd}} \sigma_{usd} \geq D_c, \quad \text{for all } c, \qquad (6.1)$$

where for all cells c, $D_c = 3$ if c is significant and $D_c = 0$ if c is insignificant. For any c, σ_{usd} leads to all stations u that can reach c (and s, d yields their respective observation field in terms of sector and distance where c is reached).

Consistency. If station u is switched on at distance $d > 1$ in sector s, it is also be switched on at distance $d - 1$ in sector s.

$$\sigma_{usd} - \sigma_{usd'} \leq 0 \quad \text{for all} \quad u, s, 1 < d \leq d_{max}, \ d' = d - 1. \tag{6.2}$$

Soft Constraints. Minimize over-coverage. Cells should not be exposed beyond their desired coverage.

$$(\textit{soft}) \ \sum_{c \in usd} \sigma_{usd} \leq D_c, \quad \text{for all } c. \tag{6.3}$$

It is important to note that in order to minimize the total over-coverage, minimizing the *number* of violated soft constraints is not sufficient. Over-coverage can occur in different degrees for each cell.

Minimizing Over-coverage in the OIP. To understand the OIP minimization problem, we first observe that it is *confined*: for every soft constraint $\mathbf{cx} \leq d$, there exists a hard constraint $\mathbf{cx} \geq d$. Therefore, the OIP minimization problem has the linear objective function

$$\min \sum_c (\sum_{c \in C_{usd}} \sigma_{usd}) - D_c, \tag{6.4}$$

and can directly be translated to an integer linear program using (6.4). The ILP can be approached with IP branch-and-bound and the linear relaxation can be used to compute lower bounds on the over-coverage.

Relation to Set-Covering. As observed above, the radar surveillance problem shares some of its structure with the set covering problem (SCP). It is the side constraints of physical consistency (6.2) that avoid the direct application of domain-specific methods for set-covering. Also, it should be noted that we are not currently aware of an *NP*-hardness result for the radar-surveillance domain.

6.1.2 Experimental Results under Varied Problem Size

This section reports on experimental results for a collection of radar surveillance instances that were generated according to different characteristics. All instances were randomly generated and vary in size (100 to 2100 cells), in the percentage of insignificant cells (0%, 2% and 5%), and in the spread of radar stations on the map (even or uneven). The density of stations remained constant. Since real placement information for radar stations was not available, insignificant cells were randomly positioned on the map. Table 6.1 summarizes the experimental results, based on a suite of radar instances generated at SICS. We ran integer local search WSAT(OIP), IP branch-and-bound CPLEX 5.0, a 0-1 simulated annealing strategy GPSIMAN [33], and the CPLEX 5.0 linear programming optimizer.

Parameters. WSAT(OIP) was run with standard parameters ($p_{hard} = 0.8$, $p_{zero} = 0.5, p_{noise} = 0.01, t = 1$) and with varying settings of Maxmoves for the different problem sizes: 30K, 100K, 300K, 500K respectively. CPLEX was run with different variations of the standard/auto parameter settings. We only report standard parameters since this yielded the overall best performance for the collection of instances. CPLEX was limited to 12h of computation time and did not reach the given memory limit of 400MB. For GPSIMAN, we used the following parameters: Maximum neighborhood size suggested by the solver (372, 776, 4447, 10772, respectively), 20 runs, 100 iterations, highest level of reoptimization, default cooling schedule (including re-heating).

Experiments with Constraint Programming. Various models (finite domain integer and Boolean) and enumeration schemes have been tried [24]. Although small problems are solved to optimality quickly, the larger sample instances could be solved with large over-coverage values only, given reasonable time. We hypothesize that it is thrashing that makes these problems hard for a constraint program that backtracks chronologically: Two distant radar stations hardly affect each other, yet with chronological backtracking the state of one station is only changed after visiting the complete subspace of configurations of many other stations.

Discussion. According to Haridi *et al.*, the long-term goal of the project is to cover a large geographical area with thousands of cells. It is thus an important criterion of success that the solution strategy scale well. The experimental results clearly show that while both IP branch-and-bound and GPSIMAN can handle small problems efficiently, problems of realistic size are beyond their size limitations.

In contrast, WSAT(OIP) is very effective on the sample problems of this domain, even for realistically sized problems. Only one class of problems with many insignificant cells (essentially 'holes' in the map) was difficult to solve. We did not systematically make attempts to improve performance on these instances because it is unknown if realistic maps would show this characteristic.[1]

From the LP optimal solutions, one observes that for many instances of the sample, the LP relaxation is tight, i.e. the optimal value of the over-coverage is the same as the LP lower bound.[2] Such problems are usually easier than problems with larger relaxation gaps (under otherwise similar parameters). The difficulty for the IP is thus closely linked with the size of the problems. Interestingly, in many cases the LP relaxation optimization takes longer than computing the optimal IP solutions using WSAT(OIP).

Dropping the Constraint Bounds. The surprising effectiveness of integer local search in this domain raises the question what the reasons are for the

---

[1] Increasing noise and decreasing p_{hard} provides some better solutions.

[2] We thank Alexander Bockmayr for initially pointing this out.

Table 6.1. Experimental comparison for radar surveillance problems: Columns are problem size in number of cells and stations (stations have a maximal reach of $d_{max} = 4$), encoding size in number of variables n and clauses m, spread of stations on the map, percentage of significant cells, and LP lower bound for over-coverage. oc* gives optimal over-coverage (integer). CPLEX 5.0 columns are: best over-coverage found, time-to-best solution, and total runtime. CPLEX was run with standard/auto parameter settings. GPSIMAN: best over-coverage found within 20 runs, mean over-coverage over all runs, and time per run. WSAT(OIP): best over-coverage found within 20 runs, mean over-coverage over all runs, and mean-time to best over-coverage over all runs. All runtimes measured on a SPARCstation 20.

size	n	m	spread	%sig	LP lb	time	oc*	CPLEX best	CPLEX to-best	CPLEX total	GPSIMAN best	GPSIMAN mean	GPSIMAN time/r	WSAT(OIP) best	WSAT(OIP) mean	WSAT(OIP) m-best
100:22	434	606	even	100	0	0s	0	opt	1s	1s	opt	0.2	10s	opt	0.0	0.0s
200:44	933	1273	even	100	1	2s	1	opt	7s	7s	opt	2.3	35s	opt	1.0	0.0s
900:200	4616	6203	even	100	2	70s	2	opt	2213s	2213s	3	12.4	528s	opt	2.0	0.6s
2100:467	10975	14644	even	100	3	391s	3	4	9.7h	12h	-	-	-	opt	3.0	1.9s
100:22	410	581	even	98	1	0s	1	opt	1s	1s	opt	1.8	9s	opt	1.0	0.0s
200:44	905	1246	even	98	2	2s	2	opt	7s	7s	opt	4.4	34s	opt	2.0	0.1s
900:200	4623	6174	even	98	4	88s	4	opt	676s	676s	17	31.3	700s	opt	5.2	4.5s
2100:467	10989	14595	even	98	11.5	661s	12	14	9.1h	12h	14	54.0	3462s	13	14.9	18.4s
100:22	371	518	uneven	100	3	0s	3	opt	1s	1s	opt	3.3	7s	opt	3.0	0.1s
200:44	772	1065	uneven	100	0	1s	0	opt	4s	4s	opt	0.8	24s	opt	0.0	0.1s
900:200	4446	5699	uneven	100	5	46s	5	6	1293s	12h	6	15.4	611s	opt	5.0	3.6s
2100:467	10771	14002	uneven	100	8.1	362s	9	11	4.2h	12h	-	-	-	10	10.8	11.7s
100:22	371	518	even	95	4	0s	4	opt	0s	0s	5	6.8	9s	opt	5.0	0.1s
200:44	772	1065	even	95	5	2s	5	opt	3s	3s	14	18.3	28s	opt	5.3	1.1s
900:200	4446	5699	even	95	19	92s	19	opt	456s	456s	61	75.3	667s	25	27.4	9.3s
2100:467	10771	14002	even	95	64.25	740s	-	67	3077s	12h	-	-	-	96	102.4	21.1s

Table 6.2. Performance drop of WSAT(OIP) when dropping constraint-bounds. STD repeats the optimal solutions from the previous table, 'no-bounds' reports on the best solution found in 20 runs, and total reports total runtime.

size	spread	%sig	oc*	WSAT(OIP)		
				STD	no-bounds	total
100:22	even	100	0	opt	opt	16s
200:44	even	100	1	opt	8	48s
900:200	even	100	2	opt	47	314s
2100:467	even	100	3	opt	93	1144s
100:22	even	98	1	opt	3	14s
200:44	even	98	2	opt	9	58s
900:200	even	98	4	opt	56	448s
2100:467	even	98	12	13	166	866s
100:22	uneven	100	3	opt	5	16s
200:44	uneven	100	0	opt	1	58s
900:200	uneven	100	5	opt	36	564s
2100:467	uneven	100	9	10	102	1132s
100:22	even	95	4	opt	10	16s
200:44	even	95	5	opt	12	58s
900:200	even	95	19	25	95	290s
2100:467	even	95	-	96	266	756s

performance. To address this question, we performed the following experiment, which is based on the hypothesis that the performance is related to the OIP problem structure. Each problem instance was modified by changing the bounds of the soft constraints from '≤ 3' to '≤ 0'. From Proposition 3.1.4 (in Section 3.1.3), we know that tightening bounds of a confined OIP does not change the set of solutions. Further, we can account for the shift in the objective function by subtracting a value from the resulting objective, $3n_s$ (if n_s is the number of insignificant cells).

Table 6.2 reports on the experimental results. Parameters were manually re-tuned to adjust for the change of the representation, resulting in switching off both the tabu mechanism and history-based tie breaking. The results in the table demonstrate that the constraint bounds are critical to obtain the previous performance. This result is consistent with our expectation because dropping the constraint bounds effectively makes the repair strategy blind with respect to which soft constraints are violated. When dropping the bounds, the search thus looses its focus and blindly makes perturbations of the variable values.

6.2 Course Assignment

The course assignment problem considered in this section deals with assigning students into pre-planned courses according to their preferences. The study was carried out based on real data of the School of Law of the University des Saarlandes in the semesters of Summer 97 and Winter 97/98, and the obtained results were used by the school to assign students to classes. As the semesters have different numbers of students, the task created a collection of real problems of varying size, ready-to-use for an investigation on real data. The problem under consideration is related to the generalized assignment problem (GAP) but includes additional side constraints.

6.2.1 Problem Description and Formulation

The scenario is the following. Students of a law school (up to 500 per semester) have to be assigned to courses (up to 30) with pre-assigned time slots and rooms. The law school offers to use a Web based interface[3] to register for a number of legal fields according the students' current interests. Further, students may submit a timetable stating the preferred time slots and the slots which they are unable to attend (*disliked* slots or aversions). As several courses are taught in each field, the aim is to assign students to courses maximizing the overall satisfaction (satisfying aversions and preferences) such that every student is assigned to one course in every one of her registered fields, while the capacity of the courses is not exceeded and the courses are not filled too sparsely.

The formulation of the problem is stated in the following, Table 6.3 summarizes the indices, constants and sets. The set F denotes the different legal fields, each field $f \in F$ is represented by a set C_f of courses. The set C contains all courses and and c_k is the desired number of participants of course $k \in C$ (usually the average number of participants of a course within the field). The fields for which a student i is registered are given by $R_i \subseteq F$. Further, the student preferences are part of the problem statement and are encoded by binary constants, $p_{ik} = 1$ if student i prefers the slot of course k and otherwise 0. Conversely, $a_{ik} = 1$ if i has dislikes course k (because k is taught during a time slot which i cannot attend).

The aim is to fill the courses within given upper and lower limits while minimizing the number of disliked assignments and minimizing the number of unsatisfied preferences (in this order). The problem can be encoded as follows: For every student i and course k, use a variable $x_{ik} = 1$ if i is assigned to k, otherwise 0 (if i is not registered for the field of course k, x_{ik} is 0).

OIP Formulation. The OIP formulation uses the following constraints.

---

[3] An application provided to the students at Saarbrücken by Reinhard Schu.

Table 6.3. Parameters for the course assignment problem.

Index	Definition
i	Index for students.
k, l	Indices for courses.
f	Index for field.

Symbol	Definition
F	Set of legal fields.
C	Set of all courses.
R_i	Fields which student i is registered for.
C_f	Set of courses in field f.
c_k	Desired number of participants of course k.
p_{ik}	binary constant, 1 iff i prefers k.
a_{ik}	binary constant, 1 iff i dislikes k.
m_i	Upper bound on satisfiable preferences for i.
u_{cap}, l_{cap}	Relative upper/lower capacity limits for courses.

Every student must attend exactly one course of each field she is registered for.

$$\sum_{k \in C_f} x_{ik} = 1, \text{ for all students } i, \text{ and fields } f \in R_i. \qquad (6.5)$$

The number of participants of a course may not exceed the desired number of participants by more than u_{cap} and not fall below l_{cap}.

$$l_{cap} \cdot c_k \leq \sum_i x_{ik} \leq u_{cap} \cdot c_k, \text{ for all courses } k \in C. \qquad (6.6)$$

No student can visit two courses that temporally overlap.

$$x_{ik} + x_{il} \leq 1, \text{ for all students } i, \text{ registered fields } s < t,$$
$$\text{such that } k \in C_s, l \in C_t, \text{ und } k, l \text{ overlap.} \qquad (6.7)$$

Soft Constraints. For every student, minimize the number of aversions, i.e. the number of assigned courses that are disliked due to their time slot.

$$(soft) \sum_{1 \leq k \leq s} a_{ik} \cdot x_{ik} \leq 0, \text{ for all students } i. \qquad (6.8)$$

Soft Constraints. For every student, minimize the number of unsatisfied preferences.

$$(soft) \sum_{1 \leq k \leq s} p_{ik} \cdot x_{ik} \geq m_i, \text{ for all students } i. \qquad (6.9)$$

where m_i is an upper bound on the number of satisfiable preferences: $m_i = \min\{|R_i|, \sum_j p_{ij}\}$ for student i where $|R_i|$ is the number of registered fields and $\sum_j p_{ij}$ is the number of preferred courses of i.

The reasoning behind the bound is as follows: Obviously, the number of registered fields is an upper bound on the number of satisfiable preferences. However, if a student has less preferred courses than registered fields, the number of satisfiable preferences is the maximum number of non-overlapping preferred courses that cover all registered fields. To keep the modeling simple, however, we approximate this value by the number of courses in preferred time slots $\sum_j p_{ij}$, which is a valid upper bound.

In order to account for the order of the goals (minimize the number of aversions first), constraints (6.8) are weighted such that constraints (6.9) are always dominated.

ILP Reducibility. In order to apply lower bounding from the direct ILP conversion, we establish confinedness first. The given OIP (6.5)–(6.8) is confined.

Proof: The soft constraints (6.8) are confined as all coefficients a_{ik} are positive. The soft constraints (6.9) are confined because the upper bounds are valid according to the above reasoning.   □

Given the confinedness of the problem, we can directly reduce it to an ILP without the need to introduce additional variables. All IP branch-and-bound experiments are subsequently applied to the transformed OIPs.

Note that in the given OIP minimization problem, the objective function value is given as a pair A–P, where A is the number of aversions in the assignment and P relates to the number of unsatisfied preferences. P may overestimate the number of unsatisfied preferences because m_i is an (approximate) upper bound on the exact number of satisfiable preferences for student i.

6.2.2 Experimental Results under Varied Problem Size

Table 6.4 reports on results of CPLEX 5.0 and WSAT(OIP). Both solvers were run with standard parameter settings. Additionally reported are CPLEX results using strong branching. Several other parameter settings have been tried (e. g. different root heuristics) but did not improve performance. CPLEX tree memory was bounded to 400 megabytes to avoid paging, runs marked [†] were cut-off to avoid paging and did not prove optimality.

Relation to the Generalized Assignment Problem. The standard generalized assignment problem can be formulated as follows. Let I be a set of agents and J be a set of jobs. For $i \in I, j \in J$, define c_{ij} as the cost of assigning job j to agent i, r_{ij} as the resource required by agent i to perform job j, and b_i as the capacity of agent i. Let x_{ij} be the binary decision variable that

Table 6.4. Course assignment, problem characteristics and experimental results. LP opt reports the optimal LP solution value, MIP lb the best MIP solution found by branch-and-bound (* provably optimal). All objective values measure the aversion–preference (A-P) values. WSAT(OIP) results report the best solution found over 20 runs, the mean best solution found, and the mean time to best solution. CPLEX * 5.0 report the best solution found, the time to the best [and second-best] solution, and the total runtime (including optimality proof if the IP optimum was found). Column CPLEX ‡ reports results obtained with the 'strong branching' strategy. All runs performed on an Intel Pentium Pro 300Mhz running Linux.

name	n	m	LP opt A-P	MIP lb A-P	WSAT(OIP)			CPLEX *			CPLEX ‡		
					best	mean	to-best	best	to-best	total	best	to-best	total
ss97-6	256	171	3-04	3-04*	opt	opt	0.0s	opt	0.0s	0.0s	opt	0.0s	0.0s
ws97-5	906	640	26-42	26-42*	opt	opt	2.1s	opt	0.1s	0.1s	opt	0.1s	0.1s
ss97-4	2288	1130	8-08	8-13*	opt	opt	0.2s	opt	145s	68m†	opt	10.1s	22.3s
ws97-3	3299	2416	2-21.8	3-24.5	3-33	3-35	6.3s	3-31	5.0s	4.9h†	3-31[33]	46m[65s]	12h
ss97-2	8404	11350	9-10	9-10.2	9-39	9-40	39.0s	14-75	47m	12.2h†	9-47[50]	12h[70m]	12h

is 1 if agent i performs job j and 0 otherwise. The *Generalized Assignment Problem* (GAP) is

$$\text{minimize} \quad \sum_{i \in I} \sum_{j \in J} c_{ij} x_{ij},$$

$$\text{subject to} \quad \sum_{i \in I} x_{ij} = 1, \qquad j \in J, \qquad \qquad \text{(GAP)}$$

$$\sum_{j \in J} r_{ij} x_{ij} \leq b_i, \quad i \in I,$$

$$x_{ij} \in \{0, 1\}$$

The course assignment problem differs from the standard GAP in the no-overlap constraints (6.7) and in the capacity constraints (6.6) that enforce a lower level of participation for every course.

Related Work on the GAP. Cattrysse and VanWassenhove [27] report that most existing techniques for the GAP are based on branch-and-bound with bounds supplied through heuristics and through relaxations of the original problem (not necessarily linear programming relaxations). According to [27], bounds are usually derived from relaxation of the assignment or capacity constraints and a variety of techniques have been applied to the GAP. Some of the existing techniques might thus be applied to the course assignment problems.

6.2.3 A Related Application: Reviewer Assignment

A domain very similar to course assignment is the assignment of reviewers in a scientific conference. For CP97[4], this method was employed to assign reviewers to conference submissions. In a bidding, all PC members stated their paper preferences. The information provided by this bidding was sufficient to assign 90% of the papers automatically. The papers with an insufficient number of bids were manually assigned by the program chair, according to the PC members' profiles and load of assigned papers. The method worked surprisingly well: In the final assignment, the load of all PC members was evenly distributed and most preferences could be satisfied.

The problem was modeled using pseudo-Boolean constraints and solved to optimality by integer local search and integer programming branch-and-bound independently.

Problem Formulation. In a bidding, all PC members submit a preference list of papers to review. Each PC member is allowed a maximum of 30 review

---

[4] The third International Conference on Principles and Practice of Constraint Programming. This section reports on joint work with Gert Smolka, program chair of CP97.

preferences in three classes: (3) *major interest*, (2) *interest*, or (1) *minor interest*. The goal is to assign as many reviewers to as many papers as possible (not more than three for each paper) while maximizing the reviewers' satisfaction and assigning a similar number of papers to all PC members. In the following, we describe a modeling of the problem using an over-constrained integer program.

The representation uses Boolean decision variables $a_{ij} = 1$ iff reviewer i reviews submission j. The biddings of the PC members are represented by the sets B_i^1, B_i^2, B_i^3 according to the preferences of PC member i (in increasing interest).

The constraints can be described informally as follows. Save resources: The number of automatically assigned reviewers for any paper is at most 3. Fair load balancing: No reviewer is to be assigned more than u papers or less than l papers (unless the bidding contains less preferences). Maximize the number of reviewers assigned to each paper. Maximize the number of class-3 paper assignments for every reviewer. And finally, minimize the number of assigned class-1 papers for every reviewer.

In the review process, the main goal is to distribute all papers. Thus the soft constraints to maximize the number of reviewers for each paper were strictly preferred over the other soft constraints by assigning them appropriate (high) weights. This enforces that papers will be distributed three times if possible without violating the hard load balancing constraints.

Results. CP97 had 132 paper submissions and thus a load of 396 reviews were to be assigned to 26 PC members. We applied three methods to solve the over-constraint pseudo-Boolean system: WSAT(OIP); a linear programming package, Loqo [145] that provided sharp lower bounds in all experiments; and a public domain integer solver, LPSOLVE, by Michel Berkelaar. For optimization, the solution quality of an assignment was computed as the summed net violation of all (weighted) soft constraints. Modeling the constraints was done from an Oz program. For the linear programming based methods, the soft constraints were translated to an equivalent modeling using an objective function. For all programs, the runtimes were under 2 seconds.

The assignment was constrained to be balanced: No PC member was to review more than $u = 20$ or less than $l = 14$ papers. The automatic assignment by this methods was well-received. On average, 86% of the class-3 preferences could be satisfied, and 32% of the class-2 preferences. Taken together with the papers with insufficient bids (that were manually assigned by the program chair) 41% of the class-1 preferences were satisfied.

6.3 Conclusions

In this chapter, two 0-1 integer optimization problems, radar surveillance covering and course assignment, have been studied, whose difficulty to a large extent is a result of their unavoidable size. For each of the problems, a confined OIP encoding was given, which was directly converted to a corresponding ILP.

Experimental results of integer local search (WSAT(OIP)) and IP branch-and-bound (CPLEX) have been reported for both domains. While for both domains, similar results were obtained for small problem instances with both frameworks, the experiments have shown that their scaling properties differ largely. Because integer local search exhibits a much more graceful scaling, WSAT(OIP) was able to outperform CPLEX by orders magnitude (runtime) on some of the largest given problems.

7. Capacitated Production Planning

"An important and widespread area of applications concerns the management and efficient use of scarce resources to increase productivity."

[Nemhauser and Wolsey, 1988]

Production planning is an important task in manufacturing systems and gives rise to a variety of optimization problems. Here we study a real-world lot-sizing problem from the process industry (manufacturing of chemicals, food, plastics, etc.). The problem is expressed as follows: given a set of products and a collection of customer orders with due dates, construct a minimal-cost production plan such that all orders are met in time without exceeding resource capacity. The total cost of a plan consists of inventory and labor costs.

The problem under consideration is similar to the well-studied capacitated lot-sizing problem (CLSP, see [43] for a survey) but includes the requirement of discrete lot-sizes that prevents a direct application of domain-specific methods from the literature [41, 95, 72]. We therefore approach the problem with a new domain-independent heuristic for integer optimization, WSAT(OIP), and empirically compare it to a commercial mixed integer programming (MIP) branch-and-bound solver (CPLEX 5.0).

This chapter describes a case study of WSAT(OIP) on a large CLSP with discrete lot-sizes and fixed charges. We compare the experimental results on real data to CPLEX applied to a tight integer programming model. We find that MIP branch-and-bound can only solve a sub-class of the CLSP with discrete lot-sizes, namely the problem where fixed charges and lot-sizes are equal. Further, WSAT(OIP) is considerably more robust than CPLEX in finding feasible solutions in limited time, in particular as the capacity constraints are tightened. With respect to production cost, both methods find solutions of similar quality. We examine fixed-capacity and varied-capacity problems. Using a Lagrangean relaxation technique we provide lower bounds that prove that the fixed-capacity problems are solved with near-optimal overall cost. We show that substantial savings can be achieved by varying capacity.

Table 7.1. Parameters for the CLSP with discrete lot-sizes and fixed charges (economic production quantities, EPQs).

Index	Definition
i	Index for items/products.
t	Index for time periods.
Symbol	Definition
L_i	Lot-size of product i.
E_i	Economic production quantity of product i.
D_{it}	Demand of product i in time period t.
T_t	Total labor units available in time period t.
R_i	Unit labor requirement for product i.
C_i	Cost of carrying product i per unit/period.
Ω_{it}	Future demand of product i starting period t.
T	Number of periods.
N	Number of items.
S	Cost per labor shift.

7.1 Capacitated Lot-Sizing

The problem under consideration can be classified as single-level, dynamic-demand capacitated lot-sizing problem (CLSP) with discrete lot-sizes and fixed charges. Given is a set of products and a number of customer orders (or forecasted demands) with due dates on a finite planning horizon. The goal is to compute a minimal-cost production plan such that all customer orders are met in time. No lateness or shortage of orders is permitted. Products (or *items*) can be produced in discrete periods of the planning horizon (weeks).

Because production consumes resources and resources have limited capacity, items often have to be produced earlier than needed and carried to the period where they are shipped. Such carrying incurs inventory cost (opportunity cost of capital and storage cost) which is one of two cost factors in the problem considered here. Solving the CLSP optimally is known to be NP-hard [19]. Table 7.1 specifies the problem parameters.

The CLSP considered here has two particularities: (i) Items can only be produced in predefined quantities (*lots*) and setup costs are compensated by economic production quantities (*EPQs*). At any time, production of item i is possible in quantities of 0 or $E_i + k \cdot L_i$, where $k \geq 0$, L_i is the lot-size and E_i is the EPQ for item i (every EPQ is a multiple of the lot-size). (ii) The only resource is labor, available in either one or two shifts in any period. The amount of available labor has an associated cost (labor availability and consumption are expressed in cost units). Thus, production cost is equal to the sum of labor and inventory costs.

In the problem, labor capacity can be varied between one and two shifts. Because less capacity enforces earlier production of items, a tradeoff exists between labor and inventory costs. Because labor costs dominate inventory costs, reducing labor is critical to substantially save costs. However, due to practical considerations it is not acceptable to have too many labor level changes; thus the number of labor level changes considered was limited to 2 in our experiments.

To optimize the overall problem, we take the approach to solve a series of capacitated lot-sizing problems with different 'labor profiles' and choose the best solution, as follows.

Labor Profiles. Labor consumption varies between items and is expressed by parameters R_i in terms of resource consumption per production of one unit of item i. In any period t, the total labor consumption is limited by T_t, available in one or two shifts. One shift incurs a per-week cost of S, two shifts incur $2S$. A labor profile thus corresponds to a set $\{(t, T_t) \mid 1 \leq t \leq T, T_t \in \{S, 2S\}\}$. Possible labor profiles are restricted to the pattern 2-shifts/1-shift/2-shifts as shown in Figure 7.1 and can be denoted by an interval $[s_1, s_2]$ referring to periods $s_1 \ldots s_2$ on one shift, and periods $1 \ldots s_1 - 1$ and $s_2 + 1 \ldots T$ on two shifts. The cost of a labor profile $[s_1, s_2]$ is thus $(T - (s_2 - s_1 + 1)) \cdot 2S + (s_2 - s_1 + 1) \cdot S$.

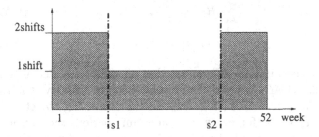

Figure 7.1. Valid labor profiles.

Every labor profile has an optimal inventory cost. If labor could be freely varied, the labor availability would have to be modeled with problem variables. However, since the number of allowed labor profiles is small, we factored the labor variability out from the optimization problem and approached the problem by solving each permitted labor profile, optimizing one CLSP at a time. Possible shift boundaries $[s_1, s_2]$ were generated starting with $s_1 = 1$ and an initial one-shift period length l ($s_2 = s_1 + l - 1$). Iteratively, s_2 was then increased as long as WSAT(OIP) found feasible solutions for the resulting CLSP (for CPLEX, as long as infeasibility was not proved). If no feasible solution was found (for CPLEX, if infeasibility of the profile was proved), s_1 was increased to the next period and s_2 was reset. The two different integer solvers require different algebraic models which are described in the following.

7.2 Integer Local Search Formulation

The integer local search model is a straightforward OIP. Production quantities per item and time period are expressed by finite domain variables p_{it} that range over the allowed production quantities (and are bounded by the summed future demand Ω_{it}):

$$p_{it} \in \{p \leq \Omega_{it} \mid p = 0 \vee p = E_i + k \cdot L\}$$

where $k = 0, 1, 2, \ldots$, for every item i and time period t and Ω_{it} is determined as $\Omega_{it} = \sum_{t < s \leq T} D_{is}$.

To formulate the constraints, we will make use of the abbreviation $S[i, t]$ representing the amount of product i carried in inventory in time period t (textually substituted in the constraints):

$$S[i, t] = \sum_{s=1}^{t} p_{is} - D_{is}$$

The OIP formulation is as follows.

$$S[i, t] \geq 0 \quad \forall i, t \qquad \text{(NOH)}$$

$$\sum_i R_i \cdot p_{it} \leq T_t \quad \forall t \qquad \text{(CAP)}$$

$$(\textit{soft}) \quad C_i \cdot S[i, t] \leq 0 \quad \forall i, t \qquad \text{(INV)}$$

Negative-on-hand constraints (NOH) ensure that all orders are met in time. Capacity constraints (CAP) express that available labor capacity is not to be exceeded. The soft constraints (INV) express the competing objectives of minimizing inventory costs; for every item and time period, the inventory cost from carrying material has to be minimized. For every feasible solution, the resulting objective (the total inventory cost) is the summed violation of all soft constraints measured by $\|.\|$ (OIP). Notice that for every soft (INV) constraint, there is a corresponding (NOH) constraint, thus the OIP is confined. Using finite domain variables to model production, the local search progresses by moving production up or down in allowed quantities induced by the violated constraints.

0-1 Integer Model. The first modeling attempt used an over-constrained 0-1 integer model with a logarithmic encoding of production quantities ($E_i x_1 + L_i x_2 + 2L_i x_3 + 4L_i x_4 + \ldots$). In addition to the blowup of the number of variables for this model, running WSAT(OIP) did not yield solutions of acceptable quality. We put this failure down to the fact that with a logarithmic encoding, a small change of production often requires a long sequence of local moves. For example, an increase from $2^k - 1$ to 2^k lots can only be achieved by flipping $k + 1$ variables. This appeared to be a strong hindrance of the search process.

7.3 Mixed Integer Programming Formulation

This section requires some familiarity with integer programming terminology, as covered for example in [114].[1] The sets and variables defined in the mixed integer programming model (MILP) are given in tables 7.1 and 7.2. The problem formulation (P) is as follows.

$$\mathbf{P} : \underset{x_{it},y_{it},z_{it},s_{it}}{\text{minimize}} \sum_{i=1}^{N} \sum_{t=1}^{T} C_i s_{it} \qquad (7.1)$$

subject to

$$x_{it} + s_{i,t-1} = D_{it} + s_{it} \quad \forall i,t \qquad (7.2)$$

$$x_{it} = L_i y_{it} \quad \forall i \in SKU_1 \qquad (7.3)$$

$$x_{it} = E_i z_{it} + L_i y_{it} \quad \forall i \in SKU_2 \qquad (7.4)$$

$$E_i z_{it} \leq x_{it} \leq \Omega_{it} z_{it} \quad \forall i \in SKU_2 \qquad (7.5)$$

$$\sum_{k=1}^{t} x_{ik} \geq L_i \lceil \sum_{k=1}^{t} D_{ik}/L_i \rceil \quad \forall i \in SKU_1, t \qquad (7.6)$$

$$\sum_{k=1}^{t-1} x_{ik} \geq \sum_{k=1}^{t-1} D_{ik} z_{it} + \sum_{k=1}^{t} D_{ik}(1 - z_{it})$$
$$\forall i \in SKU_2, t \qquad (7.7)$$

$$\sum_{i} R_i x_{it} \leq T_t \quad \forall t \qquad (7.8)$$

$$z_{it} \in \{0,1\}, y_{it} \text{ integer}$$

In the MILP model, equation (7.1) represents the sum of total inventory carrying costs. Equation (7.2) is the material balance in each time period and equations (7.3)-(7.4) determine the total production quantity of each product in time period t. Note that binary variables are only defined for $i \in SKU_2$. Equation (7.5) states that if z_{it} is non-zero, then the minimum amount (EPQ) must be produced, and cannot exceed the bound Ω_{it} (only for items in SKU_2).

Equations (7.6)-(7.7) represent constraints that tighten the relaxation gap between the integer solution and the LP relaxation of the problem. Equation (7.7) states that if product i is produced in period t, then the total amount produced up to period $t-1$ must meet the total demand up to period $t-1$. However, if the product is not made in period t, then the amount produced up to period $t-1$ must meet the demand up to period t. From our observation, this equation reduces the relaxation gap significantly and helps reduce the number of nodes branched on in a branch-and-bound solution method.

---

[1] The MILP modeling and CPLEX experiments were carried out by Ramesh Iyer and Narayan Venkatasubramanyan.

Table 7.2. Sets and decision variables for the MILP model.

	Sets
SKU	Set of products (stock keeping units).
SKU_1	Set of products for which lot-size (L_i) is equal to economic production quantity (E_i).
SKU_2	Set of products for which lot-size is a multiple of economic production quantity.
	Variables
s_{it}	Amount of product i carried in inventory in time period t.
x_{it}	Amount of product i produced in time period t.
y_{it}	Number of lots of product i produced in time period t.
z_{it}	Binary variable which is one if product i is produced in time period t.

Finally, equation (7.8) represents the labor constraints that link the problems across all products.

Due to the modeling of discontinuous integer values ($x_{it} \in \{0, E_i, E_i + L_i \ldots\}$) for items $i \in SKU_2$ with binary variables z_{it}, solving large problems is extremely expensive. We therefore attempted a Lagrangean relaxation technique (see [17] for an overview of Lagrangean relaxation) where the problem is decomposed by relaxing the equations (7.8) to obtain the value of binary variables and then solving problem (P) for fixed value of binary variables, thereby solving subproblems that are less expensive to solve in each step.

7.3.1 Lagrangean Relaxation Approach

The Lagrangean relaxation method used for solving the problem (P) relaxes the complicating constraints (7.8) using Lagrange multipliers, thus resulting in a relaxed problem that is decomposable for each i. The relaxed problem (PL) is as follows

$$\text{PL:}\underset{x_{it}, y_{it}, z_{it}, s_{it}}{\text{minimize}} \; [\sum_{i=1}^{N}\sum_{t=1}^{T} C_i s_{it}] - \sum_{t=1}^{T}\lambda_t \sum_{i=1}^{N}(R_i x_{it} - T_t),$$

subject to Equations (7.2)-(7.7).

Thus, (PL) is a relaxation of (P) and represents a lower bound to the solution of (P). Since (PL) is decomposable with respect to i, each subproblem is combinatorially less complex, and can be solved to determine the variables z_{it}. Then, for fixed values of z_{it}, the problem (P) may be solved to determine a specific solution that is an upper bound to the solution of (P).

We note that due to the discrete lot-sizes, the integer solution of (P) may result in slacks in equation (7.8) and therefore may result in all multipliers of value zero (to satisfy complementary slackness). Therefore, the multipliers λ_t for the next iteration were obtained from the LP relaxation of (P). The problem is then solved iteratively until the bounds converge. Note that the bounds are not guaranteed to converge as there may be a duality gap due to discrete nature of the problem.

7.3.2 Restricting the Problem

It is comparatively easier to solve the problem when x_{it} has no discontinuous discrete integer values. Thus, with the assumption $L_i := E_i \ \forall i \in SKU_2$, binary variables z_{it} and equations (7.5) and (7.7) can be eliminated from the formulation. Restricting a given problem instance increases the lot-sizes for all products in SKU_2, thereby reducing the set of feasible solutions. As we could not find solutions to the unrestricted problem with CPLEX, we used restricted models for all experiments with IP branch-and-bound. The restricted problem is a sub-class of the original problem.

7.4 Experimental Results

The experimental results reported in this section are based on a study of real data for 190 items and 52 weeks provided by a client of *i2 Technologies*[2] from the process industry. The OIP model resulting from the given data is large: 7520 finite domain variables (average domain size 10) and 3047 constraints (average number of variables 30, 1525 constraints soft).

To summarize the experimental results from the viewpoint of the client, what did the study achieve? (i) It found a solution which is provably within 1.4% of the optimal total cost for constant labor (two shifts), which (ii) shows that substantially cutting down cost requires reducing labor. (iii) It showed that labor can be reduced to one shift in up to 25 weeks with over 15% potential savings of total cost (or USD 1.9 million).

7.4.1 Comparison of Results

Table 7.3 reports the best solutions found by CPLEX and WSAT(OIP) in limited time and for different labor profiles. The table divides horizontally and vertically, distinguishing the original from the restricted model and the fixed-capacity from the varied-capacity case. With respect to overall quality, the best solutions among all profiles obtained from both methods are similar (WSAT(OIP) leading by less than 2%, or USD 223,295). In the experiments, the runtime of WSAT(OIP) was limited to 10 minutes, CPLEX was allowed 30 minutes for optimization and was cut-off after 60 minutes in case no feasible solution was found. All experiments were performed on a Sun Sparc Ultra II.

---

[2] A leading provider of supply chain scheduling systems.

Table 7.3. Computational results (Dollar costs) of WSAT(OIP) and CPLEX. The restricted model forces $L_i := E_i$.

cost	real problem WSAT(OIP)	CPLEX	restricted problem WSAT(OIP)
profile	fixed capacity, two shifts (230K)		
labor	11,960,000	11,960,000	11,960,000
inventory	1,023,106	1,120,680	1,040,373
total	12,983,106	13,080,680	13,000,373
profile	one shift [28,52]	[32,51]	[29,52]
labor	9,085,000	9,660,000	9,200,000
inventory	1,961,049	1,609,344	2,003,884
total	11,046,049	11,269,344	11,203,884

Run-times were kept short because many labor profiles had to be examined to find solutions of good overall quality.

Figure 7.2 and 7.3 visualize the experiments across different labor profiles for both WSAT(OIP) and CPLEX. Each impulse represents the total cost of the best solution found at one labor profile (start/size coordinates correspond to profiles [start, start+size−1], the vertical axis is overall cost). The right edge of the triangle reflects the fact that the size of the one-shift period must decrease as week 52 is approached, because the planning horizon is finite.

On the restricted model, CPLEX could not find a solution with more than 21 one-shift periods in the given time while WSAT(OIP) was able to solve a problem with 25 one-shift periods. In general, CPLEX had difficulties to find feasible solutions as the labor constraints were tightened.

Of 115 profiles solved by WSAT(OIP), CPLEX 5.0 only solved 66 profiles (57%) within the given time limit. For the profiles that could be solved with both methods, WSAT(OIP) found better solutions in 41 cases; CPLEX found better solutions in 25 cases, despite the fact that it was applied to the restricted model. In the cases where WSAT(OIP) [CPLEX] was better, on average it improved over CPLEX [WSAT(OIP)] by 2.8% [1.4%] with respect to pure inventory cost.

Parameters. CPLEX was run with standard parameter settings. Throughout the experiments with WSAT(OIP), the following parameters were used: Initial production was set to zero ($p_{zero} = 1$), and a number of 10 tries were performed, each with 100K moves. Allowed variable triggers were limited to 2 steps up or down the current variable value. Hard constraints were repaired with high priority ($p_{hard} = 0.9$). Random moves appeared to deteriorate the solution quality, therefore we set $p_{noise} = 0$. A long tabu tenure appeared to

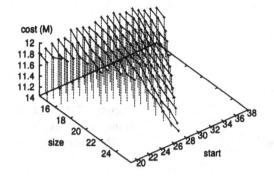

Figure 7.2. Solutions for various labor profiles (costs in USD). WSAT(OIP) on the over-constrained IP model, 10mins per profile.

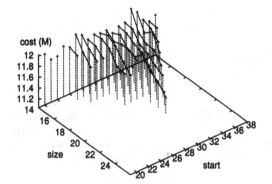

Figure 7.3. CPLEX 5.0 for various labor profiles, given the restricted MILP model, 30-60mins per profile.

be important to find feasible solutions for problems with very tight capacity ($t = 100$). Constraint weights were critical to obtain good feasible solutions and were assigned statically: The hard NOH constraints were weighted with a large number, expressing a preference to keep NOH constraints satisfied. In contrast, CAP constraints were weighted below 1.0 so that temporarily violating them during the search was encouraged.

Influence of Restricted Model. To better understand the influence of the restricted model on the solutions obtained by CPLEX, we further experimented with WSAT(OIP) on the restricted model, the results of which are plotted in Figures 7.4 and 7.5. Interestingly, WSAT(OIP) could still solve 100 of the 115 profiles given the restricted model. Crosses indicate those labor profiles for which feasible solutions exist to the restricted model that CPLEX could not find within a time bound of 1h.

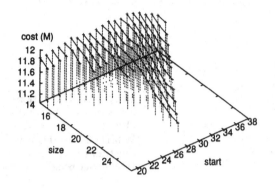

Figure 7.4. Performance of WSAT(OIP) for the restricted OIP model.

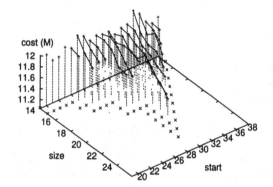

Figure 7.5. Again, CPLEX on the restricted MILP model. Crosses indicate profiles that could be solved by WSAT(OIP).

7.4.2 Lower Bounds

To assess the quality of the solutions, we applied bound reasoning based on Lagrangean relaxation as described above. We used a relaxed labor profile of constant 300K, which is over two shifts per week and therefore an unrealistic problem. For a precise estimate of the solution quality, Table 7.4 reports pure inventory costs based on this profile for the different methods. Using Lagrangean decomposition, we found solutions to the relaxed labor profile, but unfortunately could not find solutions for realistic capacity constraints. Table 7.4 also indicates that WSAT(OIP) is still considerably away from the best Lagrangean relaxation based solution (3.4% of inventory costs). With respect to the overall cost of this profile, the difference vanishes (0.2%). The reported lower bound is valid also for the original problem with constant two-shift labor, because the 300K-problem is a relaxation of the original problem.

Table 7.4. Solutions (inventory cost) based on a fixed-capacity labor profile of 300K in all weeks.

Solution/bound ($T_t = 300K$)	type	value
Best IP solution	restricted	986,780
Best solution from WSAT(OIP)	restricted	973,834
Best solution from WSAT(OIP)	original	942,511
Best Lagrangean solution	original	911,960
Best valid lower bound	original	839,875

7.5 Conclusions

We have studied a real-world capacitated lot-sizing problem (CLSP) from the process industry. Because the problem includes discrete lot-size requirements not reported in the CLSP literature, existing domain-specific methods are not directly applicable. We approached the problem with WSAT(OIP), and experimentally compared the results to a commercial mixed integer programming solver, CPLEX.

The empirical results are promising: Integer local search can solve a CLSP with discrete lot-sizes of which a commercial MIP solver can only solve a subclass. In terms of robustness, WSAT(OIP) is superior to CPLEX on the given data, in particular as the capacity constraints are tightened. The ILS model is simpler than the MIP model, and with respect to solution quality, the techniques are on par.

8. Extensions

"As fast as computers have become, they'd never be able to solve today's complex business problems without advances in algorithms ... The future, in fact, will be full of algorithms. They are moving up the complexity chain to make entire companies more efficient."

[*USA Today; December 31, 1997*]

The primary goals in this book so far have been twofold: To describe new effective algorithms within the integer local search framework and to demonstrate their capabilities for practical applications. In this section, we will critically examine the limitations of the current methods and subsequently suggest an extension to overcome some of the current limitations, and provide some ideas for future research.

8.1 Current Limitations

The limitations of integer local search as laid out here can be characterized in terms of (i) the range of problems *not* under consideration, and (ii) the factors *not* investigated by the experimental analysis.

Range of Problems. The limitations with respect to the range of studied problems are the following. First, throughout this book, we assume *pure* integer optimization problems. Practical industrial problems, however, often contain a continuous component (witness thereof is the ubiquity of the linear programming method). The framework of integer local search presented herein does not currently address mixed IPs.

Also, the case studies do not address problems in which the solutions are required to adhere to some intricate and dominating structure, such as traveling salesman problems or job-shop scheduling. Such problems require solution structure like *tours* in a traveling salesman problem or *schedules* in job-shop scheduling, which are difficult to maintain by local repairs on the level of variable assignments. For those problems, a local search in another search space (e. g. swapping cities on a tour or jobs on a critical path in a schedule) turns out to be more suitable. Alternatively if it is easy to obtain solution structure constructively, combining a greedy constructive heuristic with local moves in an abstract priority space can be very effective, as proposed in the framework of "Abstract Local Search" by Crawford, Dalal and Walser [36]. While the greedy heuristic can incorporate domain knowledge, the abstract moves change decisions in a suitable priority space ("to schedule task A earlier increase its priority").

Finally, with exception of one problem class (capacitated production planning), the problems under consideration happen to contain mainly 0-1 coefficients. In fact, for the basic version of WSAT(OIP), larger coefficients can impose a problem since the score gradient favors variables with large coefficients. Section 8.2 below suggests a way to extend the basic scoring scheme to handle large coefficients.

Experimental Analysis. Several limitations of the experimental analysis should be discussed that translate directly into suggestions for future research. First, throughout this book, performance has been studied in terms of *time to optimal solutions* and *quality of solutions obtained in limited time*. Generally, both measures of an optimization algorithm are derived from its underlying convergence behavior. To better understand the tradeoffs between time and quality, an investigation of the convergence behavior would be necessary.

Second, the proposed integer local search procedures contain several components and parameters, such as tabu search components, tie-breaking rules, noise, etc. Although some parameters have been investigated in the experimental analysis in individual case studies, a more rigorous study would be required to assess which algorithmic components and parameters are indeed critical for the success of integer local search and which parameter settings are optimal. An alternative strategy would be to strive for automatically adjusting parameters.

The case studies have demonstrated that OIP models capture the structure of many realistic problems via soft constraints, and this structure can be exploited by integer local search. Additionally, it would be interesting to consider OIP encodings of standard benchmark problems, e. g. problems from OR-library [16]. Even though many such benchmarks are randomly generated and limited with respect to their constraints, using such standard benchmarks would allow to compare domain-independent local search to domain-specific optimization heuristics from the literature, as a supplement to the comparison study using general-purpose frameworks.

8.2 An Alternative Scoring Scheme

In all of the case studies, the score as defined in (3.6) has been employed, leading to a satisfactory operational performance of WSAT(OIP). With exception to production planning, most of the arising coefficients happened be limited to $\{-1, 0, 1\}$. Preliminary experiments indicate that instances which contain larger coefficients may be more difficult for WSAT(OIP). For instance, we notice that the production planning problem requires manual weight setting. Further, some problems studied in the literature (e. g. [30]) contain larger coefficients and are not satisfactorily solved. As will be shown, this limitation is at least partly due to the applied standard score. Further, we will see that the standard score lacks motivation from a geometrical viewpoint. We will therefore propose an alternative scoring scheme next.

Let us recall the example from Section 3.4 (illustrated in Figure 3.7),

$$
\begin{array}{rll}
\text{(A)} & 9x_1 + 5x_2 & \geq 45 \\
\text{(B)} & x_1 + x_2 & \geq 6 \\
\text{(C)} & 8x_1 + 5x_2 & \leq 0 \;\; (\textit{soft}) \\
& x_1, x_2 & \in \{1, 2, \ldots, 5\}.
\end{array}
$$

We observe that at point (0,0), the contribution of constraint B to the score is 6 while A's contribution is 45. This is despite the fact that the geometric distance of point (0,0) to A and B is almost identical, and the Manhattan distance of (0,0) to A is 5, i. e. even *smaller* than the distance to B of 6.

This observation motivates reconsidering the distance function, and using the geometric Euclidean distance. From analytic geometry (e. g. [97]), we know that every inequality defines a half-space. In order to compute the distance of a point $\mathbf{p}$ to a half-space $\mathbf{ax} \geq b$ (if $\mathbf{p}$ lies outside), we compute its distance to the corresponding hyper-plane $\mathbf{ax} = b$.

To compute the distance, note that a hyper-plane can be represented in *Hessian normal form*, $(1/|\mathbf{a}|)(\langle \mathbf{a}, \mathbf{x} \rangle - b) = 0$, where $\mathbf{a}$ is a *normal vector* of the hyper-plane, $\langle \mathbf{x}, \mathbf{y} \rangle$ defines the *scalar product* $\sum_i x_i y_i$, and $|\mathbf{a}| := \sqrt{\langle \mathbf{a}, \mathbf{a} \rangle}$. The distance of a point $\mathbf{p}$ to the hyper-plane is then computed [97] as $(1/|\mathbf{a}|) \, |\langle \mathbf{a}, \mathbf{p} \rangle - b|$.

Hence, the *Euclidean distance* of a point $\mathbf{p}$ from the *half-space* defined by the constraint $\mathbf{ax} \geq b$ can be computed as

$$
\frac{\|b - \mathbf{ap}\|}{|\mathbf{a}|}, \tag{8.1}
$$

where $\|.\|$ is defined as usual (OIP), and the resulting distance is 0 if $\mathbf{p}$ lies inside the half-space.[1]

Surprisingly, we observe that for constraint $\mathbf{ax} \geq b$, (8.1) is exactly the original score weighted by $1/|\mathbf{a}|$. That is, Euclidean scoring amounts to a weighting scheme that can be computed statically as a preprocessing of the constraints.

To preserve equality between the score of a feasible solution and its objective function value, we propose to leave the soft constraints unweighted.

---

[1] Note that $\mathbf{ap}$ is the scalar product since $\mathbf{a}$ is a row vector.

Hence, the *Euclidean score* for an OIP $((\mathbf{a}_i), \mathbf{b}, C, \mathbf{d}, D)$ shall be defined as follows

$$score(\mathbf{x}) = w_{hard} \cdot \|\mathbf{b} - A\mathbf{x}\|_{\lambda^e} + \|C\mathbf{x} - \mathbf{d}\|, \qquad (8.2)$$

where the weight vector λ^e is defined as $\lambda_i^e = 1/|\mathbf{a}_i|$ for constraint $\mathbf{a}_i \mathbf{x} \geq b_i$ and $\|.\|_\lambda$ is defined as usual (3.6). Additionally, w_{hard} is a weight on the hard constraints.

For confined OIPs, one geometrically motivated way to choose w_{hard} is to base it on the corresponding objective function: If the soft constraint violation is $\|C\mathbf{x} - \mathbf{d}\|_1$ and we write this as a classic objective function $\mathbf{cx} - d :=$ $\|C\mathbf{x}-\mathbf{d}\|_1$ (which is possible as the OIP is confined), then w_{hard} can be chosen as $|\mathbf{c}|$ in order to balance the overall violation of hard and soft constraints.

Example. In example (3.12), the Euclidean score results in a weighting scheme of $w_{hard} = \sqrt{8^2 + 5^2} = 9.43$ due to the soft constraint. Further, the weights are $\lambda_A^e = 1/\sqrt{9^2 + 5^2}$, and $\lambda_B^e = 1/\sqrt{1 + 1}$. In the trajectory example, the same optimal trajectory is obtained as illustrated in Figure 3.8(c), which was obtained for a simpler, manually generated weighting scheme.

Preliminary experiments. Two preliminary experiments can be reported with Euclidean scoring. First, in the production planning case study, Euclidean scoring automatically leads to a similar weighting scheme and to similar experimental results as the scheme obtained in a tedious process of manual adjustment.

Second, to probe the effectiveness of Euclidean scoring, we tested it on a benchmark problem from generalized assignment (GAP) from OR-library [16] (GAP has been described in Section 6.2.2). We considered one problem instance from the set of 'large-sized' instances, referenced by C-1 (500 variables, 110 constraints) and described by Chu and Beasley [30].

In the experiment, we employed deterministic rounding of the linear relaxation to initialize the 0-1 variables. Using standard scoring, WSAT(OIP) could not find a feasible solution to problem C-1. When applying Euclidean scoring, an integer solution (with cost 1934) was usually found within a few seconds by WSAT(OIP), which is close to the optimal IP solution (1931) and to the LP optimum (1923.975).

It took significantly longer (many restarts) to find the IP optimum using this technique. We expect that the technique of dynamic search space reduction [9, 5] (described in Section 3.3.3) will improve on this result, considering that 341 out of 500 binary variables can be pruned after a few seconds, based on the 1934-solution. Moreover, the problem class would provide a good test case for *leashed local search* (Section 3.3.2) since the percentage of non-integral solutions of the LP optimum is very small. We are looking forward to perform an extensive experimental study of integer local search for this problem class.

8.3 Future Research

Within the endeavor to find general-purpose heuristics for combinatorial optimization, this work has established a link between local search for propositional satisfiability (SAT) and integer optimization. From this perspective, as SAT local search strategies are continuously improving, the particular strategy is insignificant compared to the possibilities that arise due to the link, encouraged by the empirical results of this first generalization. SAT can be used as a test-bed to obtain better core algorithms while integer encodings can leverage their applicability.

We discuss some of the future paths that appear most promising to pursue in order to further improve and extend the proposed methods.

Incorporating Meta-heuristics and Learning. A variety of meta-heuristic techniques have been proposed for combinatorial optimization, and sophisticated strategies have been presented for particular problem classes. In particular genetic algorithms and tabu search [61] offer a variety of strategies that would be immediately applicable in an integer local search framework, some of which are very likely to enhance the proposed strategies (e. g. tabu search intensification and more complex diversification rules, or genetic crossover).

A route that we predict will lead to very powerful integer local search solvers is the incorporation of learning strategies recently proposed for local search. Candidate strategies include reactive search [15, 14] which proposes a history-based feedback scheme, or STAGE [23] which automatically learns evaluation (scoring) functions for combinatorial optimization. Also, learning strategies for constraint-weights [132, 111, 28] or arc-weights [140] would be interesting to integrate in the WSAT(OIP) framework for further performance improvements.

Extensions of Integer Local Search. An interesting path to investigate is the connection between iterative repair and mathematical programming, which is a largely unexplored area. There will be a need to address the combination of integer local search with optimization strategies for continuous variables. Possibly, integer local search might be combined with other heuristics for integer programming to achieve this goal. Also, the formal incorporation of maximization functions into OIP will be a task to address.

With respect to the supply of constraints, we expect to have more expressive constraints available soon to extend the current expressivity of OIP. In several practically relevant cases, this can be done without the need to change the current repair strategy. For instance, within ILP it is difficult to express in the constraint $minseq(k, [x_1, \ldots, x_n]) = l$, which requires that if one of the variables x_i is assigned to l, then it must be part of a subsequence of at least k variables that are all assigned to l. This constraint can naturally be handled by the current repair mechanism.

Hierarchical Factoring. In many real-life situations, a given optimization problem is embedded into a larger context. Often, some high-level decisions *dominate* many low-level decisions because they have strong ramifications and trigger which low-level decisions need to be made. For example, in the Progressive Party Problem, deciding if a given boat is host or guest dominates the decision which boat a guest visits in a given time period. In this sense, there is a decision hierarchy in many real-life optimization problems that needs to be accounted for. Tree search approaches naturally account for such decision hierarchies by simply first making the high-level decision. In contrast, all current local search frameworks lack any support for a decision hierarchy, and it would be an interesting task to develop repair strategies that take decision hierarchies into account.

Towards a Local Search Based Constraint Solver. The class of integer linear programs covers a wide range of practically important problems and provides a good starting point for general-purpose heuristics. Nevertheless, some problems exhibit a more complex structure and need more expressive constraint representations. In contrast to integer programming frameworks that rely on linear relaxations, local search is not limited with respect to the underlying constraint systems. Hence, local search strategies $\text{WSAT}(\mathcal{X})$ are likely to appear that can handle more complex constraint systems $\mathcal{X}$, or even constraints from very different domains.

The current two-stage control strategy is (i) select a constraint c for repair, (ii) select a partial repair for constraint c. We expect that this control strategy allows for integrating a variety of more complex constraints. For instance, symbolic constraints from finite domain constraint programming (e. g. *all-different*) or constraints that address more complex structure (e. g. traveling salesman tours) will need to be integrated. Different constraints will require different strategies of local repair. The challenge of *mixed constraint systems* will be to extend the two-stage control strategy into a mature architecture that integrates different local neighborhoods and effectively control execution.

In addition to handling problems that are *larger* or *more constrained*, the challenge for local search will thus be to handle more complex problem *structure*. For instance, optimally planning the manufacturing process of a set of items by sequencing a number of ordered tasks that allocate different resources, while minimizing lateness and inventory costs. We predict that the complex structure of such real-world scenarios will be an important measure of the next generation of local search architectures.

8.4 Conclusions

In this monograph we have presented a new effective approach to domain-independent integer optimization based on generalizing local search for propositional satisfiability. The approach is applicable to a wide variety of combinatorial optimization problems that arise in practical applications. It operates on an algebraic representation similar to integer linear programs and is thus flexible and can directly be applied to realistic problem encodings.

In this *integer local search* framework, a combinatorial optimization or constraint problem is stated by an encoding with hard and soft linear constraints over finite domain variables, called an *over-constrained integer program*. The structure of this representation lends itself well to iterative repair approaches since it encodes the optimization objectives by many competing soft constraints instead using of a monolithic objective function. With respect to expressivity, we have shown that this representation is a special-case of integer linear programs.

While the local search strategy that we have presented is simple, we have empirically demonstrated its efficiency, scalability and robustness in a variety of case studies on realistic integer optimization problems. The problems either stem from the recent literature, from operating applications, or from industrial cooperation.

We have experimentally evaluated the described methods in comparisons with the literature and with general-purpose optimization strategies. The results show that integer local search outperforms or competes with state-of-the-art integer programming (IP) branch-and-bound and constraint programming (CP) approaches for the problems under consideration, in finding feasible or near-optimal solutions in limited time. The presented method, WSAT(OIP), is arguably general-purpose because neither integer programming branch-and-bound nor finite domain constraint programming can currently solve the range of problems that have all been solved with integer local search in our case studies.

A drawback of all current local search strategies is their incompleteness, that is their inability to prove infeasibility of an input problem or the quality of the achieved solutions. To partially overcome this drawback, we have discussed several effective combinations with linear programming for lower bounding, initialization by rounding, search space reduction and feasibility testing. We believe that the iterative repair strategy of integer local search offers many opportunities for improvements to the core strategy and for further generalization to more expressive constraint systems, making it applicable to structurally yet more complex problems in the future.

A. A Complete AMPL Model for ACC97/98

The following AMPL model describes the full set of constraints [115, 141] of the ACC 97/98 basketball scheduling problem which was investigated in Section 5.2.

```
##
## AMPL Model of the Atlantic Coast Competition 1997/98
## (Basketball/Sports scheduling)
## 1997-99 J.P.Walser, Programming Systems Lab, UdS

##
## Parameters
##

set Teams ordered          := { 'Clem', 'Duke', 'FSU', 'GT', 'UMD',
                                 'UNC', 'NCSt', 'UVA', 'Wake' };
set Places ordered         := Teams union { 'Bye' };
set Rounds ordered         := { 1..2*card(Teams) };
set Weekdays ordered       := { 1..last(Rounds) by 2 };
set Weekends ordered       := { 2..last(Rounds) by 2 };
set February ordered       := { 11..18} ;
set Mirror                 := { (1,8), (2,9), (3,12), (4,13), (5,14),
                                 (6,15), (7,16), (10,17), (11,18) };
param Final                := last(Rounds);
param GameQualityWeekend {Teams, Teams};
param GameQualityWeekday {Teams, Teams};

## Variables: Pairings
## T[i,j,t]=1 iff team i plays at place j in round t

var T {Teams, Places, Rounds} binary;

##
## DDRR constraints
##

subject to OP {i in Teams, t in Rounds}:
        sum { j in Places}  T[i,j,t] = 1;

subject to OV {j in Teams, t in Rounds}:
        sum {i in Teams: i<>j}  T[i,j,t] <= 1;

subject to CP {i in Teams, j in Teams, t in Rounds: i<>j}:
        T[j,j,t] - T[i,j,t] >= 0;

subject to DRR {i in Teams, j in Teams: i<>j}:
        sum {t in Rounds} T[i,j,t] = 1;
```

```
##
##  Redundant constraints
##

subject to TB {i in Teams}:
        sum {t in Rounds} T[i,'Bye',t] = 2;
subject to OB {t in Rounds}:
        sum {i in Teams} T[i,'Bye',t] = 1;
subject to HH {t in Rounds}:
        sum {i in Teams} T[i,i,t] = floor(card(Teams)/2);


##
##  Sequence constraints (Rounds)
##

# Treating Bye as Home, no more than 2 Away games
# in a row: [ #(A)<=2 ]
subject to SEQ1 {i in Teams, t in Rounds :
            t <= prev(Final,Rounds,2)}:
    sum {s in t..next(t,Rounds,2), o in Teams: o<>i} T[i,o,s] <= 2;


# Treating Bye as Away, no more than 2 Home games
# in a row: [ #(H)<=2 ]
subject to SEQ3 {i in Teams, t in Rounds :
            t <= prev(Final,Rounds,2)}:
    sum {s in t..next(t,Rounds,2)} T[i,i,s] <= 2;


# Treating Bye as Away, no more than 3 Away games
# in a row: [ #(BA)<=3 ],  (Bye+Away=not(Home))
subject to SEQ2 {i in Teams, t in Rounds :
            t <= prev(Final,Rounds,3)}:
    sum {s in t..next(t,Rounds,3)} (1-T[i,i,s]) <= 3;

# Treating Bye as Home, no more than 4 Home games
# in a row: #(BH)<=4 ]
subject to SEQ4 {i in Teams, t in Rounds :
            t <= prev(Final,Rounds,4)}:
    sum {s in t..next(t,Rounds,4)} (T[i,i,s] + T[i,'Bye',s]) <= 4;

##
##  Sequence constraints (Weekends)
##

# Treating Bye as Home, no more than 2 Away games
# in a row: [ #(A)<=2 ]
subject to SEQ1w {i in Teams, t in Weekends :
                t <= prev(last(Weekends),Weekends,2)}:
    sum {s in t..next(t,Weekends,2), o in Teams: s in Weekends
            and o<>i} T[i,o,s] <= 2;

# Treating Bye as Away, no more than 2 Home games
# in a row: [ #(H)<=2 ]
subject to SEQ3w {i in Teams, t in Weekends :
                t <= prev(last(Weekends),Weekends,2)}:
    sum {s in t..next(t,Weekends,2): s in Weekends} T[i,i,s] <= 2;
```

```
# Treating Bye as Away, no more than 3 Away games
# in a row: [ #(BA)<=3 ],  (Bye+Away=not(Home))
subject to SEQ2w {i in Teams, t in Weekends :
                  t <= prev(last(Weekends),Weekends,3)}:
    sum {s in t..next(t,Weekends,3): s in Weekends}
             (1-T[i,i,s]) <= 3;

# Treating Bye as Home, no more than 4  Home games
# in a row: #(BH)<=4 ]
subject to SEQ4w {i in Teams, t in Weekends :
                  t <= prev(last(Weekends),Weekends,3)}:
    sum {s in t..next(t,Weekends,3): s in Weekends}
             (T[i,i,s] + T[i,'Bye',s]) <= 3;


##
## Mirror constraints
##

subject to MIR1 {i in Teams, j in Teams, (s,t) in Mirror: i<>j}:
        (1-T[i,j,s]) + T[j,i,t] >= 1;
subject to MIR2 {i in Teams, (s,t) in Mirror}:
        (1-T[i,'Bye',s]) + T[i,'Bye',t] >= 1;


##
## ACC Specific Constraints
##

# No team finishes AA
subject to FAA {i in Teams}:
        sum {t in prev(Final,Rounds)..Final}
        (T[i,i,t] + T[i,'Bye',t]) >= 1;

# Of 9 weekend Rounds, each team plays 4 home, 4 on the road
# and one bye
subject to SAT1 {i in Teams}:
        sum {t in Weekends} T[i,i,t] = 4;
subject to SAT2 {i in Teams}:
        sum {t in Weekends} T[i,'Bye',t] = 1;

# Home or bye at least on two of first five weekends
subject to FIF {i in Teams}:
        sum {t in Weekends: ord(t) <= 5}
            (T[i,i,t] + T[i,'Bye',t]) >= 2;


##
## ACC/season specific constraints
##

subject to RIV: # Rival matches
        T['Duke','UNC', Final] + T['UNC', 'Duke',Final] +
        T['Clem','GT',  Final] + T['GT',  'Clem',Final] +
        T['NCSt','Wake',Final] + T['Wake','NCSt',Final] +
        T['UMD', 'UVA', Final] + T['UVA' ,'UMD', Final] >= 3;

# Popular matches in Feb
subject to FEB {(i,j) in {('Wake','UNC'),('Wake','Duke'),
               ('GT','UNC'),('GT','Duke')}}:
        sum {t in February} (T[i,j,t] + T[j,i,t]) >= 1;
```

```
# Opponent ordering constraints
subject to OPOa {i in Teams, t in Rounds: i<>'Duke' and i<>'UNC'
            and t <= prev(Final,Rounds)} :
        sum {s in t..next(t,Rounds)}
                    (T[i,'Duke',s] + T[i,'UNC',s]) <= 1;

subject to OPOb {i in Teams, t in Rounds: i<>'Duke' and i<>'UNC'
            and i<>'Wake' and t <= prev(Final,Rounds,2)}:
        sum {s in t..next(t,Rounds,2), o in {'Duke','UNC','Wake'}}
                    (T[i,o,s] + T[o,i,s]) <= 2;

##
##    Other idiosyncratic constraints
##

set FixGames dimen 4 within {Teams, Places, Rounds, 0..1};
subject to FIX {k in 0..1, (i,j,t,k) in FixGames}: T[i,j,t] = k;

data;
set FixGames :=
        # UNC plays Duke instantiated
        Duke    UNC     11      1
        UNC     Duke    18      1
        Duke    Bye     16      1
        Wake    Wake    17      0

        # as Wake is bye in Slot 1 the other must be home
        Wake    Bye     1       1
        Clem    Clem    1       1
        FSU     FSU     1       1
        GT      GT      1       1

        # duke cannot be bye here either
        Duke    Duke    18      1

        # rest
        FSU     Bye     18      0
        NCSt    Bye     18      0
        UNC     Bye     1       0

        # additions [from Trick's revisions of May 8, 1998]
        # Wake has a bye in slot 1 and must end AH
        Wake    Wake    18      1
        Wake    Bye     17      0
;
model;

subject to IDI2: T['UNC','Clem',2]+T['Clem','UNC',2] = 1;
subject to IDI3: T['Clem','Clem',Final]+T['Clem','Bye',Final] = 1;
subject to IDI5: T['UMD','UMD',Final]+T['UMD','Bye',Final] = 1;
subject to IDI6: T['Wake','Wake',Final]+T['Wake','Bye',Final] = 1;

##
##    From Trick's revisions of May 9, 1998
##

# every team must have an H in the first three slots
subject to FTH {i in Teams}:
        sum {t in 1..3} T[i,i,t] >= 1;
```

```
# every team must have an H in the last three slots
subject to LTH {i in Teams}:
        sum {t in prev(Final,Rounds,2)..Final} T[i,i,t] >= 1;



##
##  Optimization Criteria
##


##
##  Criterion 1
##  Avoid opening AA (not more than 1 team)
##  use the complement: sum over teams home or bye >= card(Teams)-1
##

subject to OAA: sum {i in Teams}
      (T[i,i,1]+T[i,'Bye',1]
    + T[i,i,2]+T[i,'Bye',2]) >= card(Teams)-1;

##
##  Criterion 2
##  Game qualities: A/B/bad Rounds
##
##  Variables: Slot-Quality
##  Each slot is either an A,B, or bad slot
##

var Q {February, 0..2} binary;
subject to SLOTQ {t in February}:  sum {q in 0..2} Q[t,q] = 1;

data;
# GameQualityWeekday[H,A]=2 means if team H plays
# home and A visits it is a 2 match (quality A)
#  A-match: 2,  B-match: 1

param GameQualityWeekday :
        Clem Duke FSU GT UMD UNC NCSt UVA Wake :=
    Clem 0 0 0 0 0 1 0 0 0
    Duke 0 0 0 1 2 0 0 1 1
    FSU  0 0 0 0 0 0 0 0 0
    GT   0 1 0 0 1 2 0 0 1
    UMD  0 2 0 1 0 2 0 1 0
    UNC  1 2 0 1 1 0 0 0 0
    NCSt 0 1 0 0 0 1 0 0 1
    UVA  0 1 0 0 0 0 0 0 1
    Wake 0 1 0 1 0 1 1 0 0 ;

param GameQualityWeekend :
        Clem Duke FSU GT UMD UNC NCSt UVA Wake :=
    Clem 0 0 0 0 0 2 0 0 0
    Duke 0 0 0 1 2 2 0 1 1
    FSU  0 0 0 0 0 0 0 0 0
    GT   0 0 0 0 1 0 0 0 1
    UMD  0 0 0 0 0 0 0 0 0
    UNC  0 0 0 1 1 0 0 0 0
    NCSt 0 1 0 0 0 1 0 0 1
    UVA  0 1 0 0 0 0 0 0 0
    Wake 0 1 0 1 0 1 1 0 0 ;
model;
```

```
# link T and Q variables: If a slot is A, there is at
# least one A or at least two B games
subject to LTQ1 {t in February: t in Weekends}:
        2*Q[t,2] <= sum {v in Teams, h in Teams: v<>h}
                            GameQualityWeekend[h,v] * T[v,h,t];

subject to LTQ2 {t in February: t in Weekdays}:
        2*Q[t,2] <= sum {v in Teams, h in Teams: v<>h}
                            GameQualityWeekday[h,v] * T[v,h,t];

# if a slot is B, there is at least one B game
subject to LTQ3 {t in February: t in Weekends}:
        1*Q[t,1] <= sum {v in Teams, h in Teams: v<>h}
                            GameQualityWeekend[h,v] * T[v,h,t];

subject to LTQ4 {t in February: t in Weekdays}:
        1*Q[t,1] <= sum {v in Teams, h in Teams: v<>h}
                            GameQualityWeekday[h,v] * T[v,h,t];


# Require 3 A Rounds in February
subject to MAXARounds: sum {t in February} Q[t,2] >= 3;

# Require <= 2 bad Rounds in February
subject to MINBADRounds: sum {t in February} Q[t,0] <= 2;

##
##   Criterion 3
##   Home/Away/Bye pattern criteria
##

set RoundsM3   ordered := { 1..last(Rounds)-2 };
set RoundsM3We ordered := { 2..last(Rounds)-2*2 by 2 };

##
## Variables expressing optimization criteria
## e.g.  for each team i and round t,
##        HB3[i,t]=1 if a sequence of at least 3
##        home games starts for team t in round i
## NOTE: "X=1 if Y" reads "Y -> X" (not iff)
##

var HB3    {Teams, RoundsM3}   binary;
var AB3    {Teams, RoundsM3}   binary;
var HB3We {Teams, RoundsM3We} binary;
var AB3We {Teams, RoundsM3We} binary;


# HB3=1 if, treating Bye as Home, 3 Home games occur in a row:
subject to HB3L {i in Teams, t in RoundsM3}:
        sum {s in t .. next(t,Rounds,2)}
          (T[i,i,s] + T[i,'Bye',s]) <= 2+HB3[i,t];

# AB3=1 if, treating Bye as Away, 3 Away games occur in a row:
subject to AB3L {i in Teams, t in RoundsM3}:
        sum {s in t .. next(t,Rounds,2)}
          (1-T[i,i,s]) <= 2+AB3[i,t];
```

```
# Similarly for weekends:
# HB3=1 if, treating Bye as Home, 3 Home games occur in a row:
subject to HB3LWe {i in Teams, t in RoundsM3We}:
        sum {s in t .. next(t,Weekends,2): s in Weekends}
            (T[i,i,s] + T[i,'Bye',s]) <= 2+HB3We[i,t];


# AB3=1 if, treating Bye as Away, 3 Away games occur in a row:
subject to AB3LWe {i in Teams, t in RoundsM3We}:
        sum {s in t .. next(t,Weekends,2): s in Weekends}
            (1-T[i,i,s]) <= 2+AB3We[i,t];


## Formulate constraints on optimization criteria
subject to HB3LE4:
    sum {i in Teams, t in RoundsM3}   HB3[i,t] <= 4;
subject to AB3LE3:
    sum {i in Teams, t in RoundsM3}   AB3[i,t] <= 3;
subject to HB3WeLE5:
    sum {i in Teams, t in RoundsM3We} HB3We[i,t] <= 5;
subject to AB3WeLE4:
    sum {i in Teams, t in RoundsM3We} AB3We[i,t] <= 4;
```

References

[1] AARTS, E., AND LENSTRA, J. K., Eds. *Local Search in Combinatorial Optimization.* Wiley-Interscience Series in Discrete Mathematics and Optimization, 1997.

[2] AARTS, E. H., KORST, J. H., AND VAN LAARHOVEN, P. J. Simulated annealing. In *Local Search in Combinatorial Optimization*, E. Aarts and J. K. Lenstra, Eds. Wiley, 1997, pp. 91–120.

[3] ABOUDI, R., AND JÖRNSTEN, K. Tabu search for general zero-one integer programs using the pivot and complement heuristic. *ORSA Journal on Computing 6*, 1 (1994), 82–93.

[4] ABRAMSON, D., DANG, H., AND KRISHNAMOORTHY, M. A comparison of two methods for solving 0–1 integer programs using a general purpose simulated annealing algorithm. *Annals of Operations Research 63* (1996), 129–150.

[5] ABRAMSON, D., AND RANDALL, M. A simulated annealing code for general integer linear programs. *Annals of Operations Research* (1998). To appear.

[6] AGGOUN, A., CHAN, D., DUFRESNE, P., FALVEY, E., GRANT, H., HEROLD, A., MACARTNEY, G., MEIER, M., MILLER, D., MUDAMBI, S., PEREZ, B., VAN ROSSUM, E., SCHIMPF, J., TSAHAGEAS, P. A., AND DE VILLENEUVE, D. H. ECL^{i}PS^e 3.5. User manual, European Computer Industry Research Centre (ECRC), Munich, Germany, Dec. 1995.

[7] ANDERSON, E. J., GLASS, C. A., AND POTTS, C. N. Machine scheduling. In *Local Search in Combinatorial Optimization*, E. Aarts and J. K. Lenstra, Eds. Wiley, 1997, pp. 361–414.

[8] APPLEGATE, D., AND COOK, W. A computational study of the job-shop scheduling problem. *ORSA Journal on Computing 3*, 2 (1991), 149–156.

[9] BALAS, E., AND MARTIN, C. Pivot and complement – a heuristic for zero-one programming. *Management Science 26* (1980), 86–96.

[10] BAPTISTE, P., PAPE, C. L., AND NUIJTEN, W. Incorporating efficient operations research algorithms in constraint-based scheduling. In *Proceedings of the first international joint workshop on Artificial Intelligence and Operations Research* (1995). Timberline Lodge, Oregon.

[11]   BARR, R. S., GOLDEN, B. L., KELLY, J. P., RESENDE, M. G., AND WILLIAM R. STEWART, J. Designing and reporting on computational experiments with heuristic methods. *Journal of Heuristics 1* (1995), 9–32.

[12]   BARTH, P. Linear 0-1 inequalities and extended clauses. Tech. Rep. MPI-I-94-216, Max-Planck Institut für Informatik, Im Stadtwald, 66123 Saarbrücken, Germany, 1994.

[13]   BARTH, P., AND BOCKMAYR, A. Modelling mixed-integer optimisation problems in constraint logic programming. Research Report MPI-I-95-2-011, Max-Planck-Institut für Informatik, Im Stadtwald, D-66123 Saarbrücken, Germany, November 1995.

[14]   BATTITI, R. Reactive search: Toward self-tuning heurisitcs. In *Modern Heuristic Search Methods*, V. Rayward-Smith, I. Osman, C. Reeves, and G. Smith, Eds. Wiley, 1996, ch. 4.

[15]   BATTITI, R., AND PROTASI, M. Reactive search, a history-sensitive heuristic for max-sat. *ACM Journal of Experimental Algorithmics* (1997).

[16]   BEASLEY, J. Or-library: distributing test problems by electronic mail. *Journal of the Operational Research Society 41*, 11 (1990), 1069–1072.

[17]   BEASLEY, J. E. Lagrangean relaxation. In *Modern Heuristic Techniques for Combinatorial Problems*, C. R. Reeves, Ed. Halsted Press, 1993, pp. 70–150.

[18]   BISSCHOP, J., AND MEERAUS, A. On the development of a general algebraic modeling system in a strategic planning environment. *Mathematical Study 20* (1982), 1–29.

[19]   BITRAN, G., AND YANASSE, H. Computational complexity of the capacitated lot size problem. *Management Science 28* (1982), 1174–1186.

[20]   BOCK, F. An algorithm for solving 'travelling-salesman' and related network optimization problems. Manuscript associated with talk presented at the Fourteenth National Meeting of the Operations Research Society of America, 897, 1958.

[21]   BOCKMAYR, A., AND KASPER, T. Branch-and-infer: A unifying framework for integer and finite domain constraint programming. *INFORMS J. Computing* (1998). To appear.

[22]   BORNING, A., FREEMAN-BENSON, B., AND WILSON, M. Constraint hierarchies. In *Over-constrained Systems*, M. B. Jampel, E. Freuder, and M. Maher, Eds. Springer, 1996.

[23]   BOYAN, J. A., AND MOORE, A. W. Learning evaluation functions for global optimization and boolean satisfiability. In *Proceedings Fifteenth National Conference on Artificial Intelligence (AAAI-98)* (1998), pp. 3–10.

[24] BRAND, P., HARIDI, S., AND OLSSON, O. Some radar surveillance problems. Tech. rep., Swedish Institute of Computer Science, SICS, 1997. To appear.

[25] CAIN, W. The computer-assisted heuristic approach used to scheduling the major league baseball clubs. In *Optimal Strategies in Sports*, S. Ladany and R. Machol, Eds., no. 5 in Studies in Management Science and Systems. North-Holland Publishing Co., 1977, pp. 32–41.

[26] CARLIER, J., AND PINSON, E. An algorithm for solving the job-shop problem. *Management Science 35*, 2 (1989), 164–176.

[27] CATRYSSE, D., AND WASSENHOVE, L. A survey of algorithms for the generalized assignment problem. *European Journal of Operational Research* (1992), 260–272.

[28] CHA, B., AND IWAMA, K. Adding new clauses for faster local search. In *Proceedings Thirteenth National Conference on Artificial Intelligence (AAAI-96)* (1996).

[29] CHA, B., IWAMA, K., KAMBAYASHI, Y., AND MIYAZAKI, S. Local search algorithms for partial maxsat. In *Proceedings AAAI-97* (1997).

[30] CHU, P., AND BEASLEY, J. A genetic algorithm for the generalised assignment problem. *Computers & Operations Research 24*, 1 (1997), 17–23.

[31] CHVÁTAL, V. *Linear Programming*. W.H. Freeman, 1983.

[32] CODOGNET, P., AND DIAZ, D. Compiling constraints in clp(FD). *Journal of Logic Programming 27*, 3 (June 1996), 185–226.

[33] CONNOLLY, D. General purpose simulated annealing. *Journal of the Operational Research Society 43* (1992), 495–505.

[34] CRAWFORD, J., AND AUTON, L. Experimental results on the crossover point in Random 3SAT. *Artificial Intelligence* (1996). To appear.

[35] CRAWFORD, J., AND BAKER, A. Experimental results on the application of satisfiability algorithms to scheduling problems. In *Proceedings AAAI-94* (1994), pp. 1092–1097.

[36] CRAWFORD, J. M., DALAL, M., AND WALSER, J. P. Abstract local search. In *Proceedings of the AIPS-98 Workshop on Planning as Combinatorial Search* (1998). In conjunction with The Fourth International Conference on Artificial Intelligence Planning Systems, AIPS-98.

[37] CROES, G. A method for solving traveling salesman problems. *Operations Research 6* (1958), 791–812.

[38] DAVENPORT, A., TSANG, E., WANG, C., AND ZHU, K. GENET: A connectionist architecture for solving constraint satisfaction problems by iterative improvement. In *Proceedings AAAI-94* (1994).

[39] DAVENPORT, A. J. *Extensions and Evaluation of GENET in Constraint Satisfaction*. PhD thesis, Department of Computer Science, University of Essex, 1997.

[40] DAVIS, M., LOGEMANN, G., AND LOVELAND, D. A machine program for theorem-proving. *Journal of the ACM 5* (1962), 394–397.

[41] DIABY, M., BAHL, H., KARWAN, M., AND ZIONTS, S. A Lagrangean relaxation approach for very-large-scale capacitated lot-sizing. *Management Science 38*, 9 (1992), 1329–1340.

[42] DINCBAS, M., HENTENRYCK, P. V., SIMONIS, H., AGGOUN, A., AND GRAF, T. The constraint logic programming language CHIP. In *Proceedings International Conference on Fifth Generation Computer Systems* (1988), Y. Kodratoff, Ed., Springer-Verlag, pp. 693–702.

[43] DREXL, A., AND KIMMS, A. Lot sizing and scheduling – survey and extensions. *European Journal of Operational Research 99* (1997), 221–235.

[44] DURBIN, R., AND WILLSHAW, D. An analogue approach to the travlling salesman problem using an elastic net method. *Nature 326* (1987), 689–691.

[45] FERLAND, J., AND FLEURENT, C. Computer aided scheduling for a sports league. *INFOR 21* (1991), 47–65.

[46] FOURER, R. A simplex algorithm for piecewise-linear programming iii: Computational analysis and applications. *Mathematical Programming 53* (1992), 213–235.

[47] FOURER, R., AND GAY, D. M. Large scale optimization: State of the art. In *Experience with a Primal Presolve Algorithm*, W. Hager, D. Hearn, and P. Pardalos, Eds. Kluwer Academic Publishers, 1994, pp. 135–154.

[48] FOURER, R., GAY, D. M., AND KERNIGHAN, B. W. A modeling language for mathematical programming. *agement Science 36* (1990), 519–554.

[49] FOURER, R., GAY, D. M., AND KERNIGHAN, B. W. *AMPL, A Modeling Language for Mathematical Programming*. Boyd & Fraser publishing Company, 1993.

[50] GAREY, M. R., AND JOHNSON, D. S. *Computers and Intractability: A Guide to the Theory of NP-completeness*. W.H. Freeman and Company, 1979.

[51] GENT, I., MACINTYRE, E., PROSSER, P., AND WALSH, T. The constrainedness of search. In *Proceedings AAAI-96* (1996).

[52] GENT, I., AND WALSH, T. An empirical analysis of search in GSAT. *Journal of Artificial Intelligence Research 1* (September 1993), 47–59.

[53] GENT, I., AND WALSH, T. Towards an understanding of hill-climbing procedures for SAT. In *Proceedings AAAI-93* (1993), pp. 28–33.

[54] GENT, I., AND WALSH, T. Unsatisfied variables in local search. In *Hybrid Problems, Hybrid Solutions (Proceedings of AISB-95)* (1995), IOS Press.

[55] GINSBERG, M., AND MCALLESTER, D. GSAT and dynamic backtracking. In *PPCP'94: Second Workshop on Principles and Practice of Constraint Programming* (Seattle, May 1994), A. Borning, Ed.

[56] GLOVER, F. Future paths for integer programming and links to artificial intelligence. In *Computer and Operations Research* (1986), vol. 13, pp. 533–549.

[57] GLOVER, F. Tabu seaerch – part I & II. *ORSA Journal on Computing 1/2*, 3/1 (1989), 190–260/4–32.

[58] GLOVER, F., AND LAGUNA, M. Tabu search. In *Modern Heuristic Techniques for Combinatorial Problems*, C. R. Reeves, Ed. Halsted Press, 1993, pp. 70–150.

[59] GLOVER, F., AND LAGUNA, M. General purpose heuristics for integer programming–part I. *Journal of Heuristics 2*, 4 (1997), 343–358.

[60] GLOVER, F., AND LAGUNA, M. General purpose heuristics for integer programming–part II. *Journal of Heuristics 3*, 2 (1997), 161–179.

[61] GLOVER, F., AND LAGUNA, M. *Tabu Search*. Kluwer Academic Publishers, 1997.

[62] GOLDBERG, D. E. *Genetic Algorithms in Search, Optimization and Machine Learning*. Addison-Wesley, 1989.

[63] GOMES, C., SELMAN, B., AND KAUTZ, H. Boosting combinatorial search through randomization. In *Proceedings Fifteenth National Conference on Artificial Intelligence (AAAI-98)* (1998).

[64] GU, J. Efficient local search for very large-scale satisfiability problems. *SIGART Bulletin 3*, 1 (1992), 8–12.

[65] HALMOS, P. R. How to write mathematics. *L'Enseignement Mathématique 16* (1970), 123–152.

[66] HAMMER, P., AND RUDEANU, S. *Boolean Methods in Operations Research and Related Areas*. Springer, 1968.

[67] HANSEN, P., AND JAUMARD, B. Algorithms for the maximum satisfiability problem. *Computing 44* (1990), 279–303.

[68] HAO, J.-K., AND DORNE, R. Empirical studies of heuristic local search for constraint solving. In *Proceedings of the Second International Conference on Principles and Practice of Constraint Programming, CP-96* (1996), pp. 194–208.

[69] HENZ, M. Scheduling a major college basketball conference—revisited. *Draft. Submitted.* (1998).

[70] HENZ, M., SMOLKA, G., AND WÜRTZ, J. Oz—a programming language for multi-agent systems. In *13th International Joint Conference on Artificial Intelligence* (Chambéry, France, 30 August–3 September 1993), R. Bajcsy, Ed., vol. 1, Morgan Kaufmann Publishers, pp. 404–409.

[71] HILLIER, F. S., AND LIEBERMANN, G. J. *Introduction to Operations Research*. McGraw-Hill, 1995.

[72] HINDI, K. Solving a CLSP by a tabu search heuristic. *Journal of the Operational Research Society 47* (1996), 151–161.

[73] HOCHBAUM, D. S., Ed. *Approximation Algorithms for NP-hard Problems*. PWS Publishing Company, 1995.

[74] HOLLAND, J. *Adaptation in Natural and Artificial Systems.* University of Michigan Press, Ann Arbor, 1975.

[75] HOOKER, J. Needed: An empirical science of algorithms. *Operations Research 42* (1994), 201–212.

[76] HOOKER, J., AND OSORIO, M. Mixed logical/linear programming. *Discrete Applied Mathematics* (1997). To appear.

[77] HOOS, H. H. Solving hard combinatorial problems with GSAT – a case study. In *Proceedings of the 20th annual german conference on artificial intelligence (KI-96)* (1996).

[78] HOPFIELD, J., AND TANK, D. 'Neural' computation of decisions in optimization problems. *Biological Cybernetics 52* (1985), 141–152.

[79] ILOG. ILOG SOLVER 3.2, User Manual. http://www.ilog.com, 1996.

[80] ILOG, CPLEX DIVISION. *Using the CPLEX Callable Library and Base System, Version 5.0,* 1997.

[81] JAFFAR, J., AND LASSEZ, J.-L. Constraint logic programming. In *Principles of Programming Languages* (1987), pp. 111–119.

[82] JAFFAR, J., AND MAHER, M. Constraint logic programming—a survey. *Journal of Logic Programming 19/20* (1994), 503–582.

[83] JAIN, R. *The Art of computer Systems Performance Analysis.* John Wiley and Sons, 1991.

[84] JAMPEL, M. B., FREUDER, E., AND MAHER, M., Eds. *Over-Constrained Systems,* vol. 1106 of *LNCS.* Springer, 1996.

[85] JIANG, Y., KAUTZ, H., AND SELMAN, B. Solving problems with hard and soft constraints using a stochastic algorithm for MAX-SAT. In *Proceedings of the First International Joint Workshop on Artificial Intelligence and Operations Research* (1995).

[86] JOHNSON, D. S. A theoretician's guide to the experimental analysis of algorithms. http://www.research.att.com/~dsj/papers/. Preliminary draft.

[87] JOHNSON, D. S. A catalog of complecity classes. In *Handbook of Theoretical Computer Science, Vol. A,* J. Van Leeuwen, Ed. Elsevier, 1990, pp. 67–161.

[88] JOHNSON, D. S. Experimental analysis of algorithms: The good, the bad, and the ugly. Invited talk at AAAI-96., 1996.

[89] JOHNSON, D. S., AND MCGEOCH, L. A. The travelling salesman problem: A case study. In *Local Search in Combinatorial Optimization,* E. Aarts and J. K. Lenstra, Eds. Wiley, 1997, pp. 215–310.

[90] JOHNSON, D. S., AND TRICK, M. A., Eds. *Cliques, coloring, and satisfiability: 2nd DIMACS implementation challenge: DIMACS workshop 1993* (Providence, RI, 1996), vol. 26 of *DIMACS series in discrete mathematics and theoretical computer science,* American Mathematical Society.

[91] KAMATH, A., KARMARKAR, N., RAMAKRISHNAN, K., AND RESENDE, M. An interior point approach to Boolean vector function synthesis. In *36th MSCAS* (1993), pp. 185–189.

[92] KARMARKAR, N. A new polynomial time algorithm for linear programming. *Combinatorica 4* (1984), 375–395.

[93] KAUTZ, H., AND SELMAN, B. Pushing the envelope: Planning, propositional logic, and stochastic search. In *Proceedings AAAI-96* (1996), pp. 1194–1201.

[94] KAUTZ, H., AND SELMAN, B. The role of domain-specific knowledge in the planning as satisfiability framework. In *Proceedings AAAI-98* (1998).

[95] KIRCA, Ö., AND KÖKTEN, M. A new heuristic approach for the multi-item dynamic lot sizing problem. *European Journal of Operational Research 75* (1994), 332–341.

[96] KIRKPATRICK, S., GELATT, C. D., AND VECCHI, M. P. Optimization by simulated annealing. *Science, Number 4598, 13 May 1983 220, 4598* (1983), 671–680.

[97] KOECHER, M. *Lineare Algebra und Analytische Geometrie*. Springer-Verlag, 1983.

[98] KUMAR, V. Algorithms for constraint-satisfaction problems: A survey. *AI Magazin 13* (1990), 32–44.

[99] LEE, J. H., FUNG LEUNG, H., AND WING WON, H. Extending e-genet for non-binary csps. In *Proceedings of the seventh International Conference on Tools with Artificial Intelligence* (1995), pp. 338–343.

[100] LEE, J. H., FUNG LEUNG, H., AND WING WON, H. Towards a more efficient stochastic constraint solver. In *Proceedings of the Second International Conference on Principles and Practice of Constraint Programming, CP-96* (1996).

[101] LI, W., BAI, S., GU, J., SELMAN, B., CRAWFORD, J., AND WANG, D. international competition and symposium on satisfiability testing. http://www.cirl.uoregon.edu/jc/beijing, March 1996.

[102] LIN, S. Computer solutions of the traveling salesman problem. *Bell System Technical Journal 44* (1965), 2245–2269.

[103] LØKKETANGEN, A., AND GLOVER, F. Tabu search for zero-one mixed integer programming with advanced level strategies and learning. *Intl. Journal of Operations and Quantitative Management 1, 2* (1995), 89–108.

[104] LØKKETANGEN, A., JÖRNSTEN, K., AND STORØY, S. Tabu search within a pivot and complement framework. *Int. Transactions on Operations Research 1, 3* (1994), 305–316.

[105] MATHIAS, E., AND WHITLEY, L. Transforming the search space with gray coding. In *IEEE Conference on Evolutionary Computation* (1994), vol. 1, pp. 513–518.

[106] McAllester, D., Selman, B., and Kautz, H. Evidence for invariants in local search. In *Proceedings Fourteenth National Conference on Artificial Intelligence (AAAI-97)* (1997).

[107] Michel, L., and Hentenryck, P. V. Localizer, a modeling language for local search. In *Proceedings of the Third International Conference on Principles and Practice of Constraint Programming, CP-97* (1997), Springer LNCS.

[108] Minton, S., Johnston, M. D., Philips, A. B., and Laird, P. Solving large-scale constraint satisfaction and scheduling problems using a heuristic repair method. In *Proceedings Eighth National Conference on Artificial Intelligence (AAAI-90)* (1990), pp. 17–24.

[109] Minton, S., Johnston, M. D., Philips, A. B., and Laird, P. Minimizing conflicts: a heuristic repair method for constraint satisfaction and scheduling problems. *Artificial Intelligence 58* (1992), 161–205.

[110] Mitchell, D., Selman, B., and Levesque, H. Hard and easy distributions of SAT problems. In *Proceedings AAAI-92* (1992), pp. 459–465.

[111] Morris, P. The breakout method for escaping from local minima. In *Proceedings Eleventh National Conference on Artificial Intelligence (AAAI-93)* (1993).

[112] Motwani, R., and Raghavan, P. *Randomized Algorithms.* Cambridge University Press, 1995.

[113] Mühlenbein, H. Genetic algorithms. In *Local Search in Combinatorial Optimization*, E. Aarts and J. K. Lenstra, Eds. Wiley, 1997, pp. 137–172.

[114] Nemhauser, G., and Wolsey, L. *Integer and Combinatorial Optimization.* Series in Discrete Mathematics and Optimization. Wiley-Intersience, 1988.

[115] Nemhauser, G. L., and Trick, M. A. Scheduling a major college basketball conference. In *Proceedings of the 2nd International Conference on the Practice And Theory of Automated Timetabling* (1997).

[116] Nonobe, K., and Ibaraki, T. A tabu search approach to the constraint satisfaction problem as a general problem solver. *European Journal of Operational Research 106*, 2-3 (April 1998).

[117] Papadimitriou, C. H., and Steiglitz, K. *Combinatorial Optimization: Algorithms and Complexity.* Prentice-Hall, New York, 1982.

[118] Parkes, A., and Walser, J. Tuning local search for satisfiability testing. In *Proceedings AAAI-96* (1996), pp. 356–362.

[119] Peterson, C., and B.Söderberg. A new method for mapping optimization problems onto neural networks. *International Journal of Neural Systems 1* (1989), 3–22.

[120] Puget, J. A C++ implementation of CLP. In *Proceedings Second Singapore International Conference on Intelligent Systems* (1994). Singapore.

[121] PUGET, J.-F. A fast algorithm for the bound consistency of alldiff constraints. In *Proceedings Fifteenth National Conference on Artificial Intelligence (AAAI-98)* (1998), pp. 359–366.

[122] RAGHAVAN, P., AND THOMPSON, C. Randomized rounding. *Combintorica 7* (1987), 365–374.

[123] RAYWARD-SMITH, V., OSMAN, I., AND REEVES, C., Eds. *Modern Heuristic Search Methods.* Wiley, 1996.

[124] REEVES, C. R. Evaluation of heuristic performance. In *Modern Heuristic Techniques for Combinatorial Problems*, C. R. Reeves, Ed. Halsted Press, 1993, ch. 3, pp. 304–315.

[125] REEVES, C. R. Genetic algorithms. In *Modern Heuristic Techniques for Combinatorial Problems*, C. R. Reeves, Ed. Halsted Press, 1993, ch. 3, pp. 304–315.

[126] REEVES, C. R., Ed. *Modern Heuristic Techniques for Combinatorial Problems.* Halsted Press, 1993.

[127] RÉGIN, J.-C. A filtering algorithm for constraints of difference in csps. In *Proceedings Twelfth National Conference on Artificial Intelligence (AAAI-94)* (1994), pp. 362–367.

[128] RESENDE, M., AND FEO, T. A GRASP for satisfiability. In *The Second DIMACS Implementation Challenge*, M. Trick, Ed., DIMACS Series on Discrete Mathematics and Theoretical Computer Science. 1995.

[129] SARASWAT, V., AND RINARD, M. Concurrent constraint programming. In *Proceedings of the 7th Annual ACM Symposium on Principles of Programming Languages* (San Francisco, CA, January 1990), pp. 232–245.

[130] SCHREUDER, J. Combinatorial apsects of construction of competition dutch professional footbal leageus. *Discrete applied mathematics 35* (1992), 301–312.

[131] SCHULTE, C. Programming constraint inference engines. In *Proceedings of the Third International Conference on Principles and Practice of Constraint Programming* (Schloß Hagenberg, Austria, Oct. 1997), G. Smolka, Ed., vol. 1330 of *Lecture Notes in Computer Science*, Springer-Verlag, pp. 519–533.

[132] SELMAN, B., AND KAUTZ, H. Domain-independent extensions to GSAT: Solving large structured satisfiability problems. In *Proceedings of IJCAI-93* (1993).

[133] SELMAN, B., AND KAUTZ, H. An empirical study of greedy local search for satisfiability testing. In *Proceedings of IJCAI-93* (1993).

[134] SELMAN, B., KAUTZ, H., AND COHEN, B. Noise strategies for improving local search. In *Proceedings AAAI-94* (1994), pp. 337–343.

[135] SELMAN, B., LEVESQUE, H., AND MITCHELL, D. A new method for solving hard satisfiability problems. In *Proceedings AAAI-92* (1992), pp. 440–446.

[136] SMITH, B., BRAILSFORD, S., HUBBARD, P., AND WILLIAMS, H. The progressive party problem: Integer linear programming and constraint programming compared. *Constraints 1* (1996), 119–138.

[137] SMOLKA, G. The Oz programming model. In *Computer Science Today*, Lecture Notes in Computer Science, vol. 1000. Springer-Verlag, Berlin, 1995, pp. 324–343.

[138] SMOLKA, G. Problem solving with constraints and programming. *ACM Computing Surveys 28*, 4es (Dec. 1996). Electronic Section.

[139] SMOLKA, G., SCHULTE, C., AND WÜRTZ, J. *Finite Domain Constraint Programming in Oz, A Tutorial.* Programming Systems Lab, German Research Center for Artificial Intelligence, Stuhlsatzenhausweg 3, D-66123 Saarbrücken, Germany, 1998. DFKI Oz 2.0 Documentation Series, http://www.ps.uni-sb.de/oz/.

[140] THORNTON, J., AND SATTAR, A. Using arc weights to improve iterative repair. In *Proceedings Fifteenth National Conference on Artificial Intelligence (AAAI-98)* (1998).

[141] TRICK, M. Modifications to the problem description of "scheduling a major college basketball conference". http://mat.gsia.cmu.edu/acc_mod.html, Mai 1998.

[142] TSANG, E. *Foundations of Constraint Satisfaction.* Academic Press, London, 1993.

[143] TSCHICHOLD, J. *Ausgewählte Aufsätze über Fragen der Gestalt des Buches und der Typographie.* Birkhäuser, 1975.

[144] VAN HENTENRYCK, P., AND DEVILLE, U. Operational semantics of constraint logic programming over finite domains. In *Programming language implementation and logic programming, PLILP'91* (1991), vol. 528 of *Springer, LNCS*.

[145] VANDERBEI, R. J. Loqo user's manual. Tech. rep., Program in Statistics & Operations Research, Princeton University, 1992.

[146] WALLACE, M. Practical applications of constraint programming. *Constraints 1* (1996), 139–168.

[147] WALLACE, R. J., AND FREUDER, E. C. Heuristic methods for over-constrained constraint satisfaction problems. In *in [84]*. Springer, 1996.

[148] WALSER, J. Retrospective analysis: Refinements of local search for satisfiability testing. Master's thesis, University of Oregon, 1995.

[149] WALSER, J. Solving linear pseudo-boolean constraint problems with local search. In *Proceedings AAAI-97* (1997).

[150] WALSER, J., IYER, R., AND VENKATASUBRAMANYAN, N. An integer local search method with application to capacitated production planning. In *Proceedings AAAI-98* (1998).

[151] WILLIAMS, C., AND HOGG, T. Exploiting the deep structure of constraint problems. *Artificial Intelligence 70* (1994), 73–117.

[152] WINSTON, W. L. *Operations Research – Applications and Algorithms.* Duxbury Press, 1994.

[153] WÜRTZ, J. *Lösen kombinatorischer Probleme mit Constraintprogrammierung in Oz*. PhD thesis, Universität des Saarlandes, Fachbereich Informatik, Saarbrücken, Germany, Jan. 1998.

[154] YANAKAKIS, M. On the approximation of maximum satisfiability. *Proceedings of the 3rd ACM-SIAM Symposium on Discrete Algorithms* (1992), 1–9.

[155] ZWEBEN, M. A framework for iterative improvement search algorithms suited for constraint satisfaction problems. Tech. Rep. RIA-90-05-03-1, NASA Ames Research Center, AI Research Branch, 1990.

Index

Lecture Notes in Artificial Intelligence (LNAI)

Lecture Notes in Computer Science